A MINER'S LIFE

David Douglass
Joel Krieger

A MINER'S LIFE

Routledge & Kegan Paul
London, Boston, Melbourne and Henley

To Emma, a pitman's daughter

First published in 1983
by Routledge & Kegan Paul plc
39 Store Street, London WC1E 7DD,
9 Park Street, Boston, Mass. 02108, USA,
296 Beaconsfield Parade, Middle Park,
Melbourne, 3206, Australia, and
Broadway House, Newtown Road,
Henley-on-Thames, Oxon RG9 1EN

Set in 10/13pt Times
and printed in Great Britain by
T. J. Press (Padstow) Ltd, Padstow, Cornwall

Library of Congress Cataloging in Publication Data

Douglass, David.
A miner's life.
Includes bibliographical references.
1. Coal-miners – England.
I. Krieger, Joel,
1951-   II. Title.
HD8039.M62G747 1983   331.7'622334'0941   82-20440

ISBN 0-7100-9473-6

CONTENTS

PREFACE

This book is the story of a contemporary coal miner. It is a story I learned during the course of several visits to England's coalfields over the last half-dozen years, visits initially undertaken as part of the field research for a dissertation about the British coal industry as an example of the problems of centralized management in a nationalized enterprise.[1]

Comprehension of the miners' lives has come more slowly than the completion of the academic project, in part due to the assumptions about work and industry which I – and, I suspect, the vast majority of us who labour above ground – take for granted, and which must be unlearned before one can make much progress in understanding life 'down the pit'.

When I first heard of men working in twelve-inch coal seams, for example, I assumed that twelve inches referred to the height of the coal seam, and that, of course, stone or waste would be extracted above or below to make a liveable working space for the collier. This is simply not the case. In 1974 men told me that they worked a full seven-and-a-quarter-hour shift in an twelve-inch space, or in eighteen inches with four inches of water. Such low seams are unusual – although two and a half to three feet would not be exceptional – but work in them they do, lying on their backs, shovelling coal transversely across their chests on to a conveyor belt. Given the typical three-shift cycle, somewhere in the northern coalfields there are men lying just that way, in the dark and damp, doing just that work, at this moment. Or they are half-crawling, half-walking toward the coalface a mile and a half away, through pools of water, with screeching cables rushing underfoot and rusting tin and bulging stone jutting down unexpectedly from the 'roof' above.

Other images of the miner below ground and above – in the village, in the garden, in the club, or for that matter in the union lodge hall – also suffer from one's intuitive disbelief or simply from the inadequacy of information provided to the outsider. We intend in *A Miner's Life* partly to remedy this situation.

The book is divided into nine short chapters. The organization of the narrative follows the rhythms, the emphasis, the turnings and intuitive associations of this miner and many miners when speaking about themselves and their underground labours. The book concerns most directly the underground life, the actual conditions of labour, working relationships among pitmen, village life and something of the daily hazards and ever-present irony and humour of life down the pit. We also explore the miners' political life, rank-and-file attitudes toward the two national strikes, toward the union and the Coal Board, and the individual miner's role in these conflicts which placed him so much at the fore.

But first, the 'we' needs some explanation. This book is more double-authored than co-authored. There are two distinct perspectives which no single narrative voice can accommodate; two points of view which we wish to preserve. I am an occasional observer, while David Douglass is a ripper* at Hatfield Main Colliery, and National Union of Mineworkers (NUM) branch official, a pitman, working at the most difficult and dangerous underground task. Thirty-four years old, he has worked fifteen years underground, studied at Ruskin College, Oxford, published several pamphlets on the life and history of the northern coalfields, and completed a university course of study at Strathclyde. Although atypical in some ways, whatever else he may be,

Douglass is a miner. No one who has ever agreed or disagreed with him would question this.

As a Geordie displaced to South Yorkshire by the pit closure programme of the late 1960s, he has followed a well-worn path of migration from the shrinking coalfields of Durham and Northumberland to newer installations in South Yorkshire and the Midlands. His impressions mirror those of thousands of other pitmen who will not be writing books.

Together, we hope to provide a doubly personal yet reliable commentary on the general conditions and life situations of miners in Britain today.

Joel Krieger

ANOTHER SIDE OF A PREFACE

The writing of this small book covers a period of time in which change is constantly taking place. Therefore, the book reflects a number of things all at the same time. We talk of the Durham coalfield, the older pits and methods of work, of ponies and thin seams. We describe family life and union life. Much of this is based upon my recollections working in that coalfield some fifteen years ago and, of course, on my contact with the Durham miners today. Joel's comments are based upon his extensive travels to a great many collieries in Durham and elsewhere in the course of his research.

I talk of my experiences in the Doncaster coalfields which span working systems and events over the last fifteen years. Some of the methods described have been replaced at Hatfield now, but they continue today in a great many other pits. Joel's travels to coalfields further south and his visits down pits in many coalfields put him in a strong position to confirm the types of work that coalminers are engaged in as we approach the twenty-first century. Many, many colliers work today in conditions little different from the nineteenth century, and those working with modern machinery below ground face a toil and a danger unthinkable in above-ground large-scale industry. It is to these conditions that we address ourselves. The veil of silence which shrouds pit work and mining must be lifted; this is what we are attempting to do in this book.

Of course, it must be said, when talking of conditions and

danger below ground, that both have improved dramatically from what they were in the old days. This improvement can be linked to the rise in the power of the NUM and its determination to stop dangerous practices. However, to contrast conditions today to what they were twenty or thirty years ago is not to suggest that we have turned the pits into a factory floor or turned mining into anything but a highly dangerous and dirty occupation. Conditions are just as we say here, and at the same time, unsafe conditions and tragedies have been reduced significantly (as has been manpower, of course). It is probably true that the nationalized mines of Britain, with a powerful and conscientious miners' union, are the safest in the world. But to say they are 'safe' by any stretch of the imagination is to say what is untrue, and it is the quietly pervasive notion that somehow 'in this day and age' life and labour underground have lost their raw and bloody edge that this book sets out to correct.

Quite a bit of the work on the book was done prior to the introduction of the new piece-rate system, the incentive agreement. This has had a profound effect on many things, but the reader will readily see where we are talking of one system and where we are discussing another. Since we started writing this book, the men at Hatfield have done me the honour of electing me their NUM branch representative and now, in June of 1982, the Doncaster miners as a whole have given me the incredible privilege of representing them to the Executive Committee of the Yorkshire Area of the NUM, so of course even with the best will in the world, I do not get below ground as often as I used to. On the other hand, when I do, or am engaged in safety inspections and dispute site inspections, I am able to see a far greater cross-section of faces, rippings, and drivages*, etc., than I ever saw before. I still am often shocked at the conditions I see and the super-human efforts of my fellow workers. It is time that the non-mining public sees something of it too.

David Douglass

1 THE CAGE

Unhurried, the miners are squatting on their haunches, sitting on a hat, or half-leaning, half-sitting against a wall. They have the last smoke. Things are getting quieter. There is the last taste of fresh air. Teams of men are breaking away all the time, one or two heading over to the cabin to pick up their lamps, having already picked up their identity checks. They hand in the number on one round tab, attach the other to the oil lamp, or to their key rings if they don't carry the 'oiler'. This is a simple safety procedure, a way of registering the men underground to catalogue their location. It is a routine which makes no attempt to conceal the daily potential for disaster. Nevertheless, if the men still feel the danger behind the ritual, they do not reveal this.

There is a sudden break and the crowd of men move to pick up their lamps. There is something rhythmical in the cap lamps*, hung casually around the shoulders at this stage, darting to and fro in the half light.

A lot of men joke and make light of it going down the pit and actually just try to change the point, not to think about it. But myself in a period about four years ago, it started really to get bad for me and I started going through agony going down there. I used to absolutely hate it and dread it, and when I got on the cage*, when it dropped away, my heart used to go like hell and I really used to be just absolutely terrified by the whole situation, especially after the Markham

Main disaster, where the brakes failed and the men crashed to the pit bottom and thirteen men were killed.

Previous to that time, I don't think I had ever thought about going down into the pit. It never crossed my mind. I always believed that these things were sort of intrinsically safe. But just after the Markham Main disaster, they fitted a new winder at Hatfield Colliery, and it was a very sporadic machine. It used to go very quickly to start with, and there were different bumps on the drum, so the rope used to fall off quicker in some places. In the period of about three weeks after it had been fitted, we walked out loads of times. There must have been a dozen strikes at the colliery over the winder. People used to almost fall off the cage when it reached bottom, just stagger out of the thing. There used to be white, ashen faces, men throwing up at pit bottom.

Of course, it's not just me. Most people don't like it – and it's visible. They go very quiet on the cage. You'll find people in there like myself, in an animated discussion just prior to, and then just actually stepping on the cage; but then as soon as I know everybody is on it – I always go quiet at that time. Other people make light and tell jokes and carry on, and it's as well they do, because at the back of your mind you're listening to them. And you're trying to get away from the thought of it, you know. And, when the cage lifts up off the keps* and you know it's going to come, at that moment it's a terrible thing.

Now, Hatfield no longer even has keps, so the cage basically hangs at the surface without the supports beneath. The keps are considered to be no longer necessary for some reason, but it will take an awful long time before we get used to being without that reassuring settling of the cage onto the keps. At least when the cage lifted off the keps you had a minute to steel yourself and take a breath. Now, however, the cage doesn't need to lift off – it just starts to drop away from its landing.

Recently the men were being wound down the shaft when

an electric circuit on the winder malfunctioned and the automatic stop came on. The men on the cage felt themselves falling, maybe twenty feet, and you can imagine the consternation in that small, packed iron box as they felt themselves hurtling down the shaft. Deputies and overmen on the chair*, some nearing collapse, others throwing up, others red in the face, others white and dripping with sweat, all confirmed that the cage had fallen away down the shaft.

On rushing over to see the management about this incident, I was assured that it was all hysterics: 'The cage has not fallen – if it had, the rope would have wound off the drum the twenty feet and the drum would have moved around, but as it happened, the pointer showing the place of stop has not moved.' Still, the men knew what they felt. To the management it was a silly incident which had given an excuse for us to come out of the pit and hope to be paid our wages. The fact that some men could not return to work for many days escaped them.

I'm not convinced of the soundness of modern machines just because they hum instead of clank, so I went over to the winding house myself, the first time I ever saw the new winder. While talking to the engineman, who was winding material down the shaft (which goes at the same speed as the men), the engine tripped out again and the emergency stop came on before my eyes. True, the drum did not revolve and the rope did not unwind, but the rope had so much give in it, and was carrying such a weight at the end of it, that being brought suddenly from some thirty miles an hour to a dead stop – the cage bounced like a ball on the end of an elastic strip. Anyone on the cage down in the blackness suddenly being stopped, then bounced down and again up, would have the very real feeling of falling away down the shaft. Standing, looking at this huge wheel spinning and droning, gave me the most unusual feeling of foreboding – and in some way, seeing it has added to my apprehension of the journey in the cage.

One may be prepared for an assault on the senses. It won't matter. One cannot escape the dread, however one seeks to combat it. Oddly, the unwitting concentration on physical discomforts helps to allay the more metaphysical ones. Carelessly buckled kneepads, for example, their straps constantly rubbing, irritate the skin. The constant twisting and crawling affronts cramped muscles, to the point that the bits of coal scratching and water endlessly dripping go virtually unnoticed. Sometimes, this onslaught so occupies the senses that nothing else remains. It is possible even to forget why and where and what could happen. After all, one is hundreds of feet underground, so deep that numbers don't matter any more, and so the men never refer to the specific depths. One is fortuitously situated within a three-foot gap maintained by a wholly unfathomable complex of mechanical contrivances – a few props combatting all the brute laws of physics. The earth is pried open for our industrial convenience. How long will it accept the indignity? Creaking, whirring, staccato clanking, the searing cut of metal against coal and stone, unremitting shadowy images trigger fears of the dark or being caught in cramped spaces. Childhood horrors scarcely recalled peer out from the unconscious. An ageing miner explained that the fears would pass. 'Don't worry,' he said, 'the first forty years are the hardest.'

You are in a permanent state of aggression down the pit and that stimulates hard work. In fact, you take out the temper on the hard work. Take just the process of walking in-bye, toward the coalface: you try to get a little bit of speed up to get it over with, and as you're walking along, things keep hitting you in the head. They bump you down, and you're at the breaking point all the time with temper. Only after a time do you stop and cool down a bit, in yourself. There are times when it would be absolutely fatal to touch a miner because the springs are so tight.

On one occasion one old lad in full flight, digging, lifting, throwing, digging, lifting, throwing and picking up quite a speed, heard his mate through grunts shout in mock self-praise and unconcernedness at the momentous heap of stone before him, 'Give me a mountain, give me a fuckin' mountain!' The old lad behind him laughed suddenly and out shot his false teeth onto the ground. His momentum of work was such that he filled them onto the shovel and shot them back into the pack hole* before he could stop himself. This, of course, brings the other toilers out in hysterical laughter, but all in an uninterrupted rhythm. Who could understand coming on a scene like this? Dense dust, heat, water, men rocking and shovelling, sweat pouring off their faces, throwing stone off their shovels, and like demented moles laughing their daft heads off. The furious pace can make it very difficult, because if you're not keeping up and the stone starts to mount up, when the man in the front stops for a minute – he doesn't actually stop for a minute, it's maybe half a minute, to have a drink of water, put in a chew of 'baccy, have a pinch of snuff – while he and the others are doing that, you're clearing away the backlog. You've just finished that backlog and it starts again. Now that goes on all day.

You can be so exhausted that a sudden stop in the work process makes you reluctant to move, lest your lack of appreciation for the stillness triggers the whole pandemonium off again. Recently, I was engaged in the task of advancing the hydraulic chocks* – advancing in line behind the cutting machines, pulling and straining to keep up, swearing at chocks too slow to keep up the pace. I was deluged in crushed coal and sprayed by leaking hydraulic blocks, the thighs aching and the heart pumping 'like the clappers'. Suddenly the face chain* stops, so the cutting machine stops. You rush like hell to pull in the remaining chocks so you're at least caught up with the process. Then you slump down just where you are.

At first you are so happy at the inactivity, you notice

nothing. Then you notice there is a big lump of coal under your left buttock, and the iron chock control is sticking in your back, and you are getting numb in your knee because your kneepad strap is too tight. But you don't move, because ten to one if you start to get comfortable, the eerie 'zong-zong-zong' of the pre-start warning will go and you're off on the race again. So you sit, grateful for 'the minute' (as they call it), even if it is uncomfortable. You can understand why.

Working in Hatfield, my latest pit, I was in the High Hazel seam, which averaged about three or three-and-a-half feet. That means you're working on your knees all day. It's also wet, so you're probably working in water as well, and you throw the stone back out of the gob* hour after hour. Your knees ache, and the water is in your kneepads, and you're rocking back and forth all the time with small bits of stone in there. It's excruciating pain. And all the little cuts you get, of course, get filled up with salty water, which doesn't help. You're working wet through, and you're actually sweating, like a pig as they say, but more like these race horses that you see when they're finished running. The sweat's just streaming off you and all over you, kneeling, or – if you're advancing girders, bullrails as they call them in Yorkshire – lifting heavy weights. There is strain on your knees when you're shovelling the stone, and it puts a lot of weight on your back and your legs, as well.

But worst of all is non-stop shovelling, kneeling forward, throwing back, kneeling forward, throwing back, hour after hour. In the older men the cartilage sometimes works itself out of the socket, and suddenly their knee joint is paralyzed with this lump. You see an old man moaning on the floor in agony, gritting his teeth, while his mate tries to push the cartilage back in again with his hands. When your knees go, you're finished on the face – you cannot work. But there's men continue like that with weak cartilage, and it's excruciating pain.

The whole job is made dozens of times worse by the presence of water, water dripping through the roof or seeping up through the floor. Either way, the man has the extra disadvantage. His hoggers* are wet through. His tools are covered with small stones and water. The water sometimes produces dermatitis, and helps to cause bet-knee, which is a heavy swelling, inflammation of the joints caused by this excess pressure. Worse even than these is what the pitmen call miner's worm – I don't know what the medical term for it is – it is found just in the pits and in the African bush, I've heard. It's caused from men shitting in their work place. Of course there are no toilets of any kind down the pit, so you have to do it just where you are. The excrement in the water develops some bacteria, which then re-enters a man's body through the pores of his skin.

Setting supports on a wet floor adds to the danger. Apart from the materials being wet and slippy, the floor is wet and slippy; and the support you think is set tight is apt to fly out. That applies to a wooden one or a metal one. The hydraulic support, when it comes down because it's wet underneath, can fly out and hit you – apart from whatever's above it to come down. So you stand a chance of being hit and buried not only by the falling rock and coal, but also by your own supports as well.

There are other disadvantages if you're working in, say, six inches of water. You cannot see the floor underneath that you're setting the thing in. You have to clear away the soft ground in order to set the prop. And, of course, this fills up with water – and so it goes on. And these aren't unusual conditions; it's common to have water down the pit, very common.

Commonplace notions about industry are inapplicable to the pit. Looked at even logically, dispassionately, the underground world seems curiouser and curiouser. To begin with, nothing is produced – rather the commodity, coal, is extracted. There are no foremen, only deputies responsible for safety and overmen to co-ordinate production. The value of these deputies and overmen is scoffed at by the miners, who act as if they are on their own and develop strange habits to discourage any intrusion: in some pits they refuse to work when any official is in view.

Unlike any terrestrial hierarchy, in the colliery there is no seniority in the normal sense: the miner nearing retirement returns to the task and pay of his youth. When his strength fails, he moves away from the coalface, working in the road-ways, for example, or driving the paddy.* In the end, he may be moved above ground, working with the new entrants who are untroubled as yet by the underground life and anxious to take their place. Together, the unformed adolescent and the men who are ageing or physically spent transport materials, shovel, and pursue the whole variety of subsidiary tasks on the surface.

In mining, seniority is only a burden. Men labour longer hours and suffer a substantial cut in wages when they are used up, and there is something of an air of exile about the removal to surface work. The mine is a world with its own logic, and the men both proudly and regretfully acknowledge the peculiarities of it.

But perhaps the most fundamental difference is that the physical plant is uncontrollably in flux. Management can never totally control production. It is maddening for them, but geology nevertheless wins, at times defeating even the bravest technological innovations. Bevercotes Colliery in Nottingham is a powerful reminder of the force of geology and – not to put too fine a point on it – the quirkiness of the underground enterprise. Even sound management strategies and more-than-generous investment are insufficient safe-guards.

The Bevercotes seams were five and a half feet high. Compared to the narrow seams of Durham mines, it was, as one management official put it, 'a Geordie's heaven'. It was the Coal Board's pet project when full-scale production began in 1967, an offshoot of the white-hot technological revolution: the 'most modern pit in Europe'. While pit ponies still laboured in a dozen collieries in Britain, the operations at Bevercotes were to be operated by remote control. A central controller sitting underground before a futuristic console of dials and gauges would direct the operation for a specific control sector of the underground installation. By pushing buttons he would regulate the remotely operated longwall faces; the underground transport of coal, men, materials, and dirt; access and egress through two mine shafts; and operation both in the coal preparation plant and in surface and underground operations. The cutting horizon of the power loading equipment – shearer loaders taking a twenty-four-inch strip of coal with each run – would be determined by a radioactive source. Power supply equipment for the face operation and machinery for the gate would be transported by a rail-mounted structure which ran along a chain conveyor, all controlled by remote signal and called a pantechnicon.

There was no end to the innovations intended for this twenty-first-century operation. Information regarding the operational state of the plant – you can hardly call a futuristic operation a pit, and the official brochure seldom used the word 'colliery' – would be transmitted by visual eight-light indicators in a mimic display unit in the control centres. Manning would be through an electronic deployment system. The position of each worker would be transmitted to the control console, deployment boards would be set up for scanning these positions, and the information reproduced on punched tapes as the basis of a time-keeping account for each collier. From the punched tape the divisional computer facility would produce the weekly payroll without further direct labour.

By the summer of 1976, things had changed dramatically. The control rooms were empty and unused, the face operations were being performed by conventional power loading methods and paid by the existing regular national agreements. As the branch secretary put it, 'Technologically, they tried to run before they could walk.' Repeatedly, one heard stories of technological buffoonery and public relations gimmickry. When a member of the royal family was on hand to witness the inaugural operation of Bevercotes's remote-control system in January, 1967, he was – so the story goes – witness to an elaborate fraud: the coal had been supplied from the production of a neighbouring colliery and placed by hand upon the remote-controlled conveyors. Bevercotes's own equipment, it seems, was not in working order. The system was constantly breaking down. The shearer, for example, operating by sensors triggered by radioactive isotopes, failed to distinguish between a thin band of dirt running through the middle of a seam and the top edge of the seam. As a result, the machine was set to the wrong height and the coaling operation was crippled. And, experimentally directed by laser beams along the face, the shearers failed to accommodate themselves to sudden changes in the position of the face, which traversed sharp inclines and shifted from side to side.

Technological difficulties were compounded by extraordinary geological misfortune. As a member of the NCB staff at Bevercotes recalled bitterly, 'Twenty-foot faults. They don't tell you about twenty-foot faults.' Beyond the very irregular wanderings of the precious band of coal among the worthless strata, the NCB also was caught unawares by the persistent cross-cutting of the coal deposits by shale. As a result, men on each coalface still in operation at some point along the face encountered a thickness of a foot or more of oil. Conditions on the face were worsened by heat as great as ninety-five degrees in the gate ends where arduous ripping operations were performed. For a time, from the original four-quadrant design for the colliery, less than one-half of a single quadrant was workable.

Bevercotes reaffirmed the truth that the Coal Board cannot control the uncontrollable: the height of the seam, the quality of the coal, faulting, the presence of foreign substances — water, methane, oil. This, at least, they share with the men in the cage: a common humility before the natural forces unfurled down the pit, and the knowledge that powers such as these may sometimes forge moments of alliance among the most mistrusting of partners.

2 THE PIT IS STILL THE PIT

The NCB works hard to embellish mining's new image. From time to time a newspaper article on the miners' newest wage demand or the Board's latest plan for coal will include a reassuring photograph. It has become a set piece: a miner, virile and handsomely tousled after a hard day at the pit, smiling vacuously. A little overworked to be sure, but cleansed by his honest labours, the young pitman has his attention focussed, apparently, on the dance that night at the miners' welfare or on the hot meal awaiting him in the canteen on his way home to the colour television. The NCB public relations department has even produced a promotional pamphlet for recruitment which features multi-coloured day-glow pits and colliers on fast motorcycles with pretty girls in tow. Among the selling points mouthed by this disco-collier after his first day down the pit is the memorable platitude: 'Unless you're a ruddy genius you've got to aim for something steady — where you're turning out something that's *really* wanted . . . like COAL!'

The miners think these efforts are foolish, but also find them a little unnerving. They anticipate the next strike and worry that NCB publicity campaigns will arrest the sympathies of potential supporters. Worse, they feel the ads deny the miner his claim to continuity with the past. Without that sense of continuity, work down the pit would be meaningless, so the collier insists on his connection to history all the more passionately, the more he feels the continuity threatened or his place in the tradition derided. History preserves his dignity, so the claim to history becomes a personal thing and a tone of reverence casts its spell over the parade of events, some lived and all remembered.

In the decade from 1958 to 1968, 120 collieries were closed in the north-east. Manpower in Durham, for example, fell from 97,924 in 1958 to 44,160 a decade later as a series of concentrated coastal pits replaced the age-old pattern of the village colliery scattered throughout the country. In the face of this onslaught, a miner forced south by the closures explained to me the collier's sense of the pit. Like a peasant's connection with the land, it is unremitting. 'There will always be a picture of the ground underneath,' he explained. 'The pit is gone. The heap might go soon, and there'd be no remnant, but I would know it by underground still, and know where there is a shaft, and know that there is a belt of coal. The old miners tell us, so we remember where the coalfields were. They were down *here*, they were *there*, they were *over there*. We know the whole region, not simply the surface buildings, not simply the people who live there, but also what it's like underneath. We know it and that can't be destroyed.' But there is always the fear that Coal Board and technology will win, and that the past of kneepads and vests and pit sense will become a hollow metaphor, a lament of old age.

Between the pitmen and the Board, therefore, wages are in some sense a secondary matter. The most scarring battles are over the meaning of the industry – about whether the present can be severed from the past. Is the contemporary faceworker a collier or a machine operator? When the collieries are closed and the men moved into spanking new cosmopolitan installations, can they still claim their pride of place beside the big hewer? There can be no settlement in this dispute between history and progress, and no resolution in a strike. This battle about meaning is the enduring source of bitterness between management and men, and the bitterness it leaves is the surest proof that the industry hasn't changed that much.

Historically, there is a lot of superstition in the pit, particularly in the Durham and Northumberland pits, which are the oldest ones. And there is one superstition which I've always kept myself. On day shift, early morning quiet, you sit for a while. There's a special chair, the same chair every day, where you gather your thoughts while you have a cup of tea. And having left the house, you never go back. If you forget something, you don't go back. If you go back, you don't go to work. You never break that tradition. But there is a little safety net – so that if you did *have* to go back, you can re-establish this cycle by sitting down in the seat again. I try to rationalize this tradition, because it's not stupid – many of these so-called superstitions are rooted in serious events. The man attunes himself for work that day, and if something actually goes wrong to disrupt that, it's an omen that maybe in himself something's not working right. Maybe he's not quite in tune with the whole thing at all. For years and years he's done the same thing, and for one day he doesn't do it. So he thinks, 'That's it. Wouldn't it just be fate for me to go back into the house after being too bad to go to work in the first place, and then go down the pit and get injured. That would just be my fate.' And if it is something you have no control over – if there is somebody who happens to be there, and you cannot sit in your proper seat – it makes you very uncomfortable.

There's also superstition about the bottom deck of the cage. I know a lot of Geordies about the same age as me who won't ride the bottom deck on a Monday morning. We say, jokingly, it's because you get to the bottom quicker. But how many times in the old days, say a hundred and fifty years ago, did it happen that a group of accidents occurred on the day shift, with all the men killed on the bottom deck and not on the top deck? For a long time afterwards, people would say,

'I know it's just coincidence, but none the less, I'll go on the top deck.' I know in fact there've been times when they've made an extra draw for us, even when there's been room on the bottom deck. We wouldn't go down. They've had to bring the cage up again, and let us go in the top deck. They were furious about that kind of thing, but we say, 'Well, that's it, otherwise we'll go home.' And we would, too.

Just before I went away to go to university, I was forced to travel on the bottom deck and I started to think, 'Oh God, this is it. It's going to happen. I won't get away.' The fact that it didn't happen doesn't prove it isn't so and it doesn't make it feel any easier the next time. It's still the same. Then, curiously enough, I discovered that in the Markham Main cage accident it was all the men in the bottom deck who died outright, while on the top they survived, even though they were mangled into unrecognizable shapes.

Of course, part of it is that you never give an inch, and you don't concede things like that. Somebody might call it being bloodyminded, but it's a matter of keeping your dignity and hanging on, because the old issues always come back. One of the outstanding disputes we've just had at Hatfield concerns the whole issue of winding time. It's bad enough on days, but the management have tried to make our unpaid time even greater. After the 1926 strike the hours of labour were increased to include in the working hours of the shift the time it takes one way to get to the pit bottom in the cage. That winding time can be at the beginning or the end of the shift. So, if our day shift starts at 6 a.m. and the management says that everyone must be down the pit by 6 a.m. and that paid time starts then, then at the end of the seven-and-a quarter hour shift all the men must be out of the mine by a quarter past one. That is seven-and-a-quarter hours plus his winding time he had at the beginning of the shift to get the men down by 6 a.m.

Well, for quite a while the management had allowed all the deputies and overmen and management to come right to the front of the queue at lowse* and ride up *before* the men. The men might all be standing there and along come all the officials and ride up the pit, maybe two-and-a-half cagefuls of them. We always hated this but we put up with it. Suddenly, however, the gaffer* started to have a purge against men going down the pit what he called 'late', so he said that anyone not down the pit by five minutes to six had to be sent home. Imagine! Men might be standing at the shaft side at ten minutes to six after having been out of their beds at four-thirty that morning. Because they couldn't all cram on to the last cage at five minutes to six, that was it. Get off home!

We said that if you are going to have this winding time at the beginning of the shift, then all of our men must be up the pit, outside by a quarter past one. Any time later was our time, and the officials had no right therefore to ride up in front of us on our own time. We instructed our onsetters* at the bottom of the shaft not to allow officials to ride before us and to make them stand in the queue. The deputies refused, the onsetter refused therefore to allow anyone on the cage, so the manager retaliated by sending our onsetters out of the pit. Obviously we couldn't allow our men to ride with non-NUM onsetters so that was it, all NUM men were withdrawn from the pit and we were out, and the negotiations have gone to national level and still not been resolved.

We termed it a lock-out since we turned up for work every day but were not going to be allowed to work with normal agreements. We were out three days until our area agent persuaded us to go back to work pending negotiations. All this time and at this moment we are still being forced underground from five thirty in the morning and not being paid till 6 a.m. If you can get down. We are still being kept down the pit at the end of the shift up until a quarter to two after our pay stops at a quarter past one. Worse, we still have to stand in line like dummies while all the officials push their way to the front and ride up the pit before us. People think this kind of treatment which the miners have endured for centuries is a thing of the past, but as we see it, it is still very much with us. Outdated privilege is still rubbed in our face. How much have things changed?

A few years back on the occasion of the centenary of the Durham Miners' Association, a programme was issued showing the miner of yesterday hewing in his cavern. The idea was to contrast him with the immaculate conception of a space-age miner before the banks of TV screens, directing the automatic operations of a remote-operated face. Countless times I have seen the miner dressed in his pit gear leading the parade, shorts, vest, kneepads, oil lamp, helmet and boots, an exact copy in fact of what most of us do wear down the pit.The exasperating thing is when he is announced as 'an old-fashioned miner'.

The pit is still the pit, though, and a great many of us still work on our knees and our bellies. Many of us still work with shovels, so the Board disguises the true conditions when they just talk about a 'machine-cut face'. With stables* or neuks* it is true that the vast majority of coal is cut by machinery, so when the Coal Board presents the total cut by machines as against that won by hand, the figures are overwhelming and support the impression of automation. However, while virtually all the coal is cut by machinery, the number of men on a stable face involved in the cutting operation is only nine: one chargeman, one machine operator, five chocking, ramming and grading,* and two on the tail gate* stable machine. This leaves sixteen others: four men in the main gate* stable shovelling the bulk of the day, a borer involved in hand-held boring, six men on the tail gate shovelling, and six on the main gate shovelling. So, out of twenty-six men on that coalface, seventeen will be shovelling, blasting, boring, hammering home supports and toiling in much the same fashion as on a conventional handfilled face, while nine will be involved in cutting and advancing hydraulic chocks. This is so even though all the glossy Coal Board graphs of tons of machine-got coal against hand-got coal look like the vertical against the horizontal.

Although actual handfilled* faces are a minority, they are by no means extinct. In September 1979, there were still twenty-three handfilled faces, and there are an equal number that have mechanized coal cutting but conventional hand-set timber and supports. And, of course, to the men involved on those faces going into these conditions every day, it doesn't matter at all that the majority of their comrades do not have to suffer likewise.

Besides, modern machine faces are no gardens of Eden. We aren't talking about shop floor conditions or anything like it. We are speaking of a dark, hot, sometimes wet, sometimes small gap in the earth, shaking all the time. Everything is on the move, and then there's the noise and commotion and thick clouds of choking dust. In the gates, there will probably be a mechanical 'bucket' (like a small bulldozer), but this means the rippers need only spend half their shift shovelling, and they lose two men off the job to compensate for the bucket. This isn't the picture the Coal Board tries to paint. Take the idea about thick seams. There is a widespread belief that big seams are safe. In fact, in so many cases they are a nightmare every bit as bad as thin seams. Just where this myth about big seams originated is hard to tell. Perhaps thick seams represent progress and modern industry, while thin seams are seen as prehistoric. That the thin seam is horrific is not open to question, but the thick seam is hardly less so. Men conditioned to working narrow and medium seams will face agonies in high seams trying to work upright all day. It places immense burdens on the spine, which in older and even middle-aged men are never fully overcome or adapted to. The weight of everything increases tenfold, everything is bigger, bulkier and more awkward. The height of the coal itself provides a special problem from big chunks of the seam falling on to the travelling side of the face and killing people. Similarly in the gate and on the face—the height of everything ensures that even a relatively small piece of stone falling from the roof will pick up terrific momentum and cause serious injury, whereas in a small seam the same stone would cause only a graze.

Another hazard found in both thick and thin seams are the angles, tilts and rolls which plague men in their work. Although there are some faces in which the coal seam is more or less horizontal, this is by no means a general rule, and the working of faces which have steep gradients is not an uncommon feature of pit life. Some of the gradients at which coal is worked are quite dramatic. One of the deaths on the face in 1978 was on a face with a gradient of 1 in 1.8!

So how is this different from the 'old-fashioned' miner? Some of the dangers and hazards are still there as they always have been. On Sunday 18 March 1979, at Golborne Colliery, Lancashire, ten men were blown to kingdom-come by our old enemy fire damp.* Just four years before that at Houghton Main Colliery, Yorkshire, a similar accident killed five men. Vigilance about methane amongst officials is becoming so casual as to be negligent in some cases. A number of officials cannot be bothered with the flame lamp,* which is the only sure and handy method of detecting methane. It is certain that some of them cannot actually read the lamp, but are pushed through their gas tests by soft-hearted examiners who know that failure of the gas test means loss of position by the official. For this reason, many officials never ever use the oiler, and some are so criminal as to even leave it hidden out-bye. Others rely on the mechanical methanometer* which is far from failsafe, sometimes defective, and in any case, is often left out-bye with the oiler.

As with the Bentley disaster, the whole parade of concern and enquiries and recommendations is so much whitewash. This is why we get so many look-alike accidents occurring one after the other, with big disasters followed by months of investigations and recommendations which are promptly ignored. After the Golborne explosion, the British Society for Social Responsibility in Science observed:

> If there were detectors, why wasn't the seam either exhausted of gas or sealed off before repairs were begun? It is known that when a fan breaks down for a few hours gas accumulates. . . . Did the management follow the recommendations of the Houghton Main inquiry?

The pitmen were even more to the point. A Golborne miner was heard to say: 'They don't check for gas, it holds up the job. The methanometers don't get taken down the mine. It happens hundreds of times.'

Now it is made a thousand times worse by the return of wages linked to production. Of recent years, the Incentive Bonus Scheme has blinded the men to a point where many of them can see nothing but money, and they have themselves become apathetic to gas and gas detection. Very few men carry the oil lamp themselves, and of those who do, very few dare to hold up production operations even for minutes at the beginning of the shift while they test the area for gas. We are being bribed, threatened and corrupted into putting our own lives and the lives of our comrades on the line for the sake of production.

They talk about dignity, but these are the depths to which the Coal Board will sink. Here is what the Society for Social Responsibility also reported:

> The morning after the mens' arrival at Withington Hospital in South Manchester, a Registrar arrived at the Burns Unit – for which he had no clinical responsibility whatsoever – equipped with a tape recorder. A 'testimony' was extracted from one of the now dead workers, whilst he was still under the influence of

numerous drugs and delirious from pain. The Registrar
ignored the dying man's pleas for something to kill the
pain, that is until he had given his account of the
explosion. Those tapes are now in the hands of the
NCB, being used to oppose the miners' compensation
claims.

Some of the dangers and indignities are not so dramatic.
Noise is a hazard which is only just being talked about and
then very rarely and, ironically enough, very quietly. There is
next to *no* action against it. Every day brand-new equipment
is produced and sold at lavish prices to the Board, equipment
which is likely to make all who use it deaf, and also render
them defenceless against the earth by robbing them of their
highly tuned 'pit sense' and sharp ears for the moving of the
roof and crack of the coal. Without these subtle warnings the
miner is easy victim to the pit. A great many machines are
hazards in this way – boring machines, scraper chains, and
most of the compressed-air-powered earth movers.

During a survey at three mines completed in 1977, it was
found that 13 per cent of the underground men were exposed
to an equivalent continuous sound level of 90 decibels and
over, and 20 per cent of the surface workers were exposed to
comparable dangers at least part of the time. The conspiracy
on noise is one that throws its nets far wider than the pits. It
affects millions upon millions of workers. In the north of
England where heavy industry predominates, the sound
tracks on pictures are turned up louder than those down
south because it is known that most of the audience will be
industrial workers and therefore already partially deaf.

Another area the Coal Board ignores is sanitation underground. Here it is a lie by simple omission. A couple of years
back, following the pitman's holiday abroad, a big dysentery
scare started. Dysentery is very infectious, and bad sanitation
is its breeding bed. TV channels and radio were used to track
the man down 'before it was too late'. They were obviously

terrified of a huge epidemic in the mining industry. We were faced with a Medical Officer of Health on the local TV stations explaining that 'toilet facilities underground were not all they should be'. Of course every miner in the audience nearly developed diarrhoea through laughing. There are *no* toilet facilities in the pits. So what do we do? Well, if you wish to relieve yourself, you do it just where you are. There's no night soil men down the pit, so that unwanted human debris of yesterday stays with you today and every day for years every time you pass that spot. It's there when the back-bye workers clear up floor lift to lay track, or unload material, and of course where you sit and have your bait.

Twenty minutes we squat in a corner somewhere to eat dust-covered sandwiches. With the incentive scheme, bait is even shorter. Although the belts still stop for twenty minutes (sometimes even this is fiddled so the job never stops), the work never does. Maintenance work is carried on, panning over and timbering goes on, and the work on the rippings goes on without respite. The most we get now is a couple of the lads squatting down to eat a couple of slices while their mates continue boring, working with the pneumatic drills and hammers or shovelling. We don't get time now to cool down, so our hearts are still beating fit to burst and the sweat is still streaming from us as we eat. We don't even have twenty dust-free minutes in the whole shift. The bait is washed down by cold water, and not only does the Coal Board not supply the water underground – you have to carry every drop yourself – but you have to buy the bottles to carry it in. They cost you 70p from the pit canteen.

Of course, you can sometimes make extra pay. For working in water up to your knees you get fifty-five pence a day; you get one pound if it's falling down off the roof onto your head; or if it meets in the middle and drowns you, you get the two together! We also get ten pence a day for carrying an oil lamp, and around nine pence for carrying the bag of powder in and eighteen pence for two bags of powder. If you bring a wheelbarrow full, I don't think they pay you for that. But them's the payments that you get. Most of that money goes on food, clothes, rent, etc. with very little left over for enjoying oneself. There are few miners and their wives who can afford to go out drinking other than at the weekend, and holidays abroad are still an exception rather than the rule.

When all the pits were closed in the 1960s, all these people on the TV, all these ministers — people who didn't know what we know — all kept saying, 'That's not dignified work, anyway. It's just as well to run down the industry.' It's odd, now they aren't saying that. Now the Coal Board are telling us it's very dignified to go down in the pit, it's a modern industry. But the pit is still the pit. No, we didn't want the pits open. We don't go in the pits because we like the pit. We go into the pits because that's the only place, because there's money there, because that's where you earn your living. Because there is no alternative. If the alternative is rotting or starving, you've got to go to the pit. When they closed the pits we weren't bothered about that. It was the destruction of the community, and the destruction of our livelihood that we objected to. It's not really possible to see past the pit. During the big boom in the nuclear era, they closed the pits, and atomic energy was going to be the end of it all. And one old pitman heard all this and he said, 'Aye, but when they've used up all the atomic energy, they'll have to come back to the coal!' Of course, the big atomic future never happened. But at the time, what he said gave me a vision of a spaceship, like the *Great Eastern*, five funnels, and men stripped to the waist, throwing coal into great furnaces and giant paddle wheels spinning silently in space.

3 RIPPING

Traditionally, the faceworker was paid by some variation of piece-rate: for cutting by the square or cubic yard; for hauling by the hundredweight or tub. The rate of payment was set by local contract negotiated on the spot between miner or union official and the undermanager. Pay depended directly upon output. Adverse conditions which made the work more difficult and the output (hence the wage) low – a fault, for example, or water on the face – forced constant renegotiations of the agreements. Managers and men haggled daily.

The miner struggled to improve his pay through hasty work. He wasn't paid much for 'timbering-up' so the roof might be left inadquately supported. The hewer who had to wait for a tub to fill off his coal would lose that much time in his race for the wage, so two hewers might fight over the use of a tub.

Between 1966 and 1971 a uniform daily wage replaced the system for local – even individual – payment by results. For a dozen years the National Power Loading Agreement (NPLA) set the industry on a new modern footing. Every miner doing comparable work on any coalface in Britain would receive equal pay. For the first time a miner would know in advance the content of his weekly wage packet. It could be affected by neither geology nor machine breakdown. His own effort would not be a factor at the end of the week.

Then, in a major reversal in the winter of 1978, productivity bonuses were introduced. Once again, the details of the wage package would be hammered out locally, with national agreements setting the fall-back rates for particular classes of work. In principle, method study would determine standard performance, and men would be rewarded for productivity accordingly. Once again, what the miner took home at the end of the week would vary and be subject to all the vicissitudes of underground labour. The new agreement was made over the objections of the most militant voices in the NUM and among the rank and file, but with the consent of the men at large, who anticipated higher earnings and supported the plan in area-by-area ballots.

Wage systems are cyclical and always controversial – but in different ways. Piece-rate is undignified and divisive for the colliers; national time-based wages unify the mineworkers and can result – as it did in 1972 and 1974 – in national work stoppages; productivity bonus schemes pit the geographical areas and particular collieries with thick seams and high-technology methods against the less fortunate. The Board recognizes the value of this kind of division, but doesn't want to pay out bonus wages on coal it doesn't need mined or reward labour 'excessively' when unemployment makes mining, suddenly, a more attractive pursuit. They therefore seek to 'rationalize production' in other ways. This has meant the introduction of more and more supervision over the miners' work, and the reduction of all colliers and all classes of labour at the coalface to a single status. Skilled miners become *task-workers*. In this, the NCB resembles any management. They would like the mines to work as factories. Miners resent the attempt and say it is senseless and bound to fail. Pitmen think of themselves as hewers*, putters*, pullers*, caunchmen* – not as multi-skilled colliers, which is to say, machine operators.

Reasonably enough, the NCB promotes the image of a clean, streamlined, almost clinical industry in which the men are overburdened by memories of the primitive past. The

Board has gone to some remarkable lengths to disguise the nature of the industry. For example, the NCB now encloses the towering gantries and huge winding gears which symbolize the old-fashioned pit, and have tried to make the sites look more like factory works. It has introduced a 'work wear' scheme, issuing the colliers bright orange overalls so the men stop looking like their images in TV plays from the 1930s. It has covered over the walkways from the mine to the pit baths so that no one will chance to see black and bloody colliers trooping in from work. All this to demonstrate that mining isn't really mining any more and nobody has it that tough these days.

In the mid-1970s the Doncaster area production manager for the Coal Board mentioned in a television interview that working in the pits was no worse than being in the London underground. Nevertheless, any and every photo taken down the pit must be approved by the management, television cameras must be licensed, and NCB officials must approve the exact time and location of filming. Meanwhile, the television epidemic of historical plays about mining seems to render the present somehow false, as if miners who still crawl and shovel and sweat and cough black are shamming it, perversely choosing to live a life that's an anachronism when they could just as easily enjoy the simple pleasures of modern technology and progressive management.

Oddly enough, none of this – at least, before the effects of the incentive scheme held sway – affects the pace of the work. I've often wondered about it. There are some jobs that were introduced after the abolition of piece-rates, new jobs which could be performed very quickly, but they weren't. They were performed at an ordinary rate, a sensible rate. But the jobs that had been left over from piece-rates were still performed at a certain speed, even though nobody told you to. Now, although I was on the face during piece-rates, I wasn't on piece-rates myself because I had only just started at

Wardley. Even so, all of us who worked with a shovel on the face were taught at that speed, and that's the speed you worked at. So automatically, when you picked up the shovel, you worked at that speed with the shovel. And when you are knocking props in you do it at that speed. If you were building packs you do it at that speed. You didn't draw it out because it was a day wage – it's a different thing. It's almost as if the job had an intrinsic speed in itself. It's a legacy of the thing. It's like that with ripping.

Now if you are ripping, the first thing you'll do is to build a scaffold. A scaffold is made from iron bars and thick wooden bulks, and is done in order to get at the lip, the actual stone face. You then bore six to twelve – maybe twenty – holes in the lip. Boring the lip can be very strenuous. Most of the boring machines I've worked with have been windy ones, compressed air. Normally, two men assist in the boring operation, because at an unguarded moment you can hit a hard bit of stone when you are boring and the machine can whip around, and with that it can break your arm, or it can knock you unconscious if it hits you in the head. When the lip is fired down – the shot-firer fires it down – this leaves you at the bottom, with about sixteen to twenty tons of stone. And up above you a huge cavity. Then we pluck the lip, and that's going to be very dangerous, but somebody has to do it. We use a pinch-bar which is a thing like a harpoon made of iron. All you have to do is knock all the lumps of stone down with it; you stand on the scaffold and knock them until they're loose. This can take some time and of course you have to be ready to hit the thing. You have to build your scaffold far enough forward to be able to reach the lumps. But not so far forward that the stone's going to hit the scaffold, 'cause it would smash it. Though sometimes this happens, anyway. The stone could fall on you, of course. As soon as you give one stone a tap, the whole lot comes in. And you've really got to move, then. Because if it breaks the scaffold, then you've got all the iron bars and everything flying everywhere as well as the stone, as well as the pinch-bar.

Sometimes the bar can seem to follow you. It was my misfortune to be working on the right hand side of the rip shovelling when the damn bar pointed itself right at me and the moving face chain pushed it rapidly towards me – it was caught under the chain. I jumped fast off my knees and started to walk backwards down the gate pursued by the bar, which looked determined to pierce me about throat level. It's amazing how you can watch impending obliteration and yet nothing in the brain department seems to respond. The feet are walking backwards against the advancing spear, but there was no real reaction.

Suddenly the brain box clocked in and I dived down low. The bar struck into the wall of the tunnel where I had been standing. It bent like a banana then snapped; one piece carried on down the chain and the other stuck at least six inches into the stone side of the tunnel. When I emerged, I saw that my marras* had got themselves down from the bow* where they had been hanging. It was a shock for us all, but whereas their legs shook a little, my Adam's apple had trouble staying in place for a number of days. So you have to be very quick. Even then, you know you can keep hitting and hitting the roof when you're plucking it, but sometimes you've got to say, 'Well, that's as good as we're going to get it.' And it might still be falling. It's at times like this when you get marks on your back. Because if small bits fall on you, or even quite large bits fall on you, you have to stand there with this bow. And if you've got water teeming on top of you as well, you cannot look up. Your inclination is to look up and see what the top's doing, but you get your eyes full of water and you can't see what you're doing, 'cause it's this stinging chemical water.

So you lift the bow straight up on to the horseheads*, that's it – once that's up, there's a little bit of security. It's still not covered though, the lumps can still come and knock the thing off. Then not only do you get the scaffold, you get the bow and the lumps on top of you! Then you have to get it into line so the tunnel is going straight, then you get your wooden boards over the top, your lagging, or your corrugated iron sheets. Now you're getting somewhere. You've got some cover behind you, on top of you. But then the job comes of supporting, from those sheets up to the exposed ground. You have to pin the top so it's secure, and build a chock or some kind of structure with wood. It's like building a roof support on top of a roof support. Having done that you've got some safety. Then at the sides, it depends on how good a borer you are, what the lip's like. Sometimes the sides don't blast very good. So then you have to get the windy pick and drill the stone away in order to get the legs on. That's basically that.

Or at least it ought to be, but the problems aren't simply physical or geological. It has been my misfortune from time to time to encounter the pragmatic, 'no nonsense', macho, racialist, hard-fighting, big-working pitman. It would be a lie to suggest that you can subject a body of people over many decades to the conditions described here and not produce a few like this. They in their turn have had the misfortune to be stuck with a small, vegetarian, socialist, anti-sexist, anti-racialist who looked like he couldn't pull the skin off a rice pudding. Only work and crucifying slog will balance the difference between the two elements. But there comes a time when blood and every ounce of sweat in one's body will never suffice to disprove that 'small means weak'.

I remember the conclusion to a long-running war of this kind. I had been with a character like the one described above. It was his determination to discredit me through lack of work, and my determination to match him blow for blow in everything he did. It ended in the pit bottom, in front of all the men riding up off day shift – a public and bitter polemic.

'Tell me – who is the biggest, you or me?'

'Me,' he shouted back.

'Who is the tallest, you or me?'

'Me, you silly bastard.'

'Who is the heaviest, me or you?'

'Me, you daft little bastard.'

'So then when I go and carry that fourteen-by-twelve heavy section ring with you – who is working the hardest? *You* or *me*?'

And in honesty – amidst cheers from the men who love such verbal blood lettings as much as the physical ones – he replied, '*You*, you clever little bastard.'

No, that didn't end the war, though. Some people will never allow respect to be earned.

4 BLUE GILLETTES

The 'big hewer' is gone, they say with regret, the miner who would work as hard as ten ordinary men and perform wondrous feats of strength, all the time blessed with colossal 'pit sense' that saved his comrades from disasters, interpreting the creaks and moans, knowing the strata, representing the collective strength and cunning of all in the worst of times. But there is also relief. Mechanization erodes traditions and pride, but also reduces – or at least regulates – effort.

Often, ageing miners exude a paternal affection for the younger, brasher pitmen who would rather go to the pub than to the union lodge hall. They say they are smarter, more in tune with the world of modern technologies and management strategies. And who can resent the technological progress that dignifies some of the work?

Yet, the older miners don't always feel easy about the tally of benefits and losses in the modernizing process they can't control. They mourn a change of attitude – less sense of accomplishment, a fraying solidarity. 'An old miner's eyes would boggle if he came back and saw what's coming off the Panzer (a flexible armoured conveyor, introduced after the war),' explains a northern pitman, a traditionalist with mixed emotions about all this. 'The slugger and the slugger mentality are gone. The best sluggers are those Panzers.' This is the same man who spoke bitterly of the disappointment of nationalization, the hopes for a 'Shangri-La' dashed, when the old gaffers, the owners and their henchmen, returned in new hats as NCB managers. To the men, nationalization has proved a cruel disappointment, and the NCB – perhaps beyond its due – is blamed for all that goes wrong and stays awful down the pit.

One of the NCB dust samplers going down the pit to take the readings was caught going out of the pit an hour early. When asked where he was going, he said, 'I'm not going to sit in that dust!' But the dust isn't the only cause of danger and disease. It's also the material and equipment which come to the pit bottom every day.

It was common knowledge among rippers working in the gates that the wooden lagging was the best. It was thick but light. It held the weight quite well. When backripping*, it was easy to remove, or at least relatively easy, so the faceman preferred them because of size and manoeuverability. However, it is expensive, we are told, so management introduced tin sheets which are cheaper. The tin sheets may be cheaper, but they are also awkward to handle – they don't hold the weight of the gate and then they burst. They stick down into the travelling road as lethal blades. These sheets are the same as you sometimes see in surface jobs, they are corrugated, about four feet long and three feet wide. Underground, because of the confined space and other bad conditions, they pose a real problem. We set them between the arches that we use when building the tunnel. It can be a real problem getting them in (and later the backrippers might have to try and extract them again when they are worn and burst), when you don't have room to manoeuvre, and often they are very, very sharp and the hands and arms very soon become covered in cuts from the sheets. And if all this wasn't enough, certain firms, either because of pressure put upon them for lowering costs or from straightforward corruption, cut the size of the sheets. These sheets are cut down to razor thinness.

They look good, but they're useless against actual weight

and deadly to handle or walk into. Then you also discover that these wafer-thin sheets, which were once galvanized against rust, were later left ungalvanized and instead, to save money, were coated with oil. The cut arteries and heads are fine examples of the concern the NCB had for safety at the time.

When the first 'blue Gillettes' arrived in our unit, we refused point blank to work with them. Eventually all the shifts in our unit refused to use them, which held up production. Finally, the message got through to the management, who sent us some new galvanized sheets. But the old ones we'd refused were sent off to another unit to use them until, in quick succession, several men had cut arteries and deep gouges, and they refused to use them, too. The oily sheets were taken out of the pit, dried, and then sent back down again. In two days, the sheets were a mass of rusty, rotten tin, no thicker than bean cans. Now, of course, there are huge falls in the road, through the roadways, as the paper-thin sheets give way under the weight of the earth. This in turn leaves a band of rusty iron sticking down the travel road just above eye level. These are the sheets that were used and cleared by management, and the safety office.

Lack of material is an all-too-common problem underground. The deputy might report lack of timber every day, the NUM local reps on the consultative committee with management might raise it at every meeting, yet still holes cannot be timbered and the roof is left unsupported for want of boards, pinners*, or splits. Shoddy materials are also common, which is almost worse than nothing. The safety officer is paid direct by the NCB and comes under the colliery manager's orders. Safety officers see questions like this about the sheets as bloody-minded attempts by the men to stop the job and still get paid. The officers are fearful of being used as a stick to beat the manager, so in many cases they do not assert themselves as they should. The safety officers stand back and watch tram loads of potentially dangerous materials going underground. They pass timber yards full of it, they watch lorry loads go past their windows, but they remain silent. Production and keeping the peace seem to be the priorities on their list.

A number of years back at Hatfield, the pit was not doing too well in production. The response of the management was to cut down in the size of the timber, the supports, all set by hand. Chocks, planks, lids* and pinners were all reduced in size. Half tree trunks were made quarters, which of course rendered them useless, cut too thin to be of any value. The management considered that because they weren't getting as much coal out of the pit, they could cut down on the expense of proper supports. As if holes in the ground are conscious of profit ratios!

Another example of dangerous equipment is the haulage chain. Most of the men on the face are involved in driving the coal-cutting machine or advancing the face supports, timbering the broken ground, etc. These men are usually recognizable by their lack of fingers. It's a thing you notice in any pit community, even among young lads. One of the chief reasons for this is the hauler chain. The hauler chain runs the full length of the face, tight, on a huge compressed spring. It's on this chain that the coal-cutting machine pulls itself forward. And when this chain breaks, as it often does, it whiplashes down the face catching people in their heads or their bodies or else, more commonly, crushing their fingers against the sides of the chocks. This is fortunately a declining occurrence, with the introduction of the 'rackatrack' system, which does away with the hauler chain altogether.

Worse, sometimes chocks are left back for repair and the machine cuts past their place before they are brought back in. When this happens all the weight is thrown on to the tail gate lip. Often it makes the pack holes collapse and produce huge cavities in the top and sides of the gate. It therefore exposes the ripper to terrible danger from falling rocks and gas build-ups. Not until the whole tail gate end collapses does the big machine stop cutting, only by then it is too late to save the men from danger.

The same is true of the chocking teams. When weight comes on to the chocks, they often cannot be advanced immediately. This results in big areas of exposed roof, which often causes falls of ground and exposes the chocking teams to danger from falling coal and rocks whilst they timber up the cavities, and from that dreaded haulage chain.

On top of all this, we are constantly harassed about the amount of materials we use. We are urged, told and forced to use less and less. The cost of each board is rammed down our throats in the hope that we will not use them as often! As if we set all this stuff for the hell of it, and not for our safety! After the process of harassment, if we still continue to set 'too much' timber, the management have replied by simply refusing to send us any, at the same time forcing the face forward to more production, leaving vast areas of ground unsupported and in a dangerous state. In this type of operation the deputies and overmen either turn a blind eye – or else they put in report after report demanding material but to no avail.

As for this matter of being 'supervised' by deputies and overmen – underground miners have *always* done the work themselves, have never been told what to do. And the job has always been left to the miners to do. Only in recent times, since the abolition of piece-rates – especially before the new bonus schemes – have they tried to bring in a kind of factory supervision which has failed. It failed because miners will not accept supervision of that kind. Even though we're told we will work under close supervision – it says that in some of the agreements – this actually cannot happen, because the official is incapable of supervising a man twenty years his senior, and the only expertise in the operation comes from experience.

Besides, they want it both ways. They want to be able to tell you what to do at particular times, but if anything goes wrong, they want us to fix it. If, for example, there's a cave-in on the face, it won't be the gaffers – the deputies, these supervisors – who get up into the hole and timber that up – we have to do that. And we have to do that to our satisfaction, because it is us who are going to be working there. The irony is, if anybody gets killed or injured in that situation, it's not the supervisors you hear about: the management always says, 'Skilled men, they knew what they were doing.' For example, suppose we've got the tunnel going along. We blast enough ground out to be able to put an extension of the tunnel in but, because of a weak roof, or the nature of the rock, instead of blasting out fifteen feet, maybe thirty-five feet will come. So then, you have to support the residue of that, you have to set your arch and extend the tunnel, but also support all that excess ground above that has fallen out. And, of course, all the time you're doing that, it continues to fall, or it can continue to fall. More might come any time, and skittle the whole thing.

In a dangerous situation like that, the best thing a deputy can do is pass us the supplies along or to hand us the timber up. In situations like that, they stay well clear. Occasionally an ex-ripper who has become a deputy may try and help or get up to assist, but more often than not this would be resented as a slight on the skill of the men doing the job, a suggestion that they ain't doing it right. So usually we're the people who have to do it. Often, after you'd fired, when you had a big fall of ground, there'd be a gigantic hole, maybe fifteen feet over the top of the arches. Somebody has to go up there and timber it up. You can't ignore it, it won't go away. There is nothing you can do at all except get up on top of those rings with nothing above you, nothing holding the roof up but sheer will power. You think to yourself, 'Right, up you go.' You urge yourself all the time, 'Come on ye bugger, come on,' and curse the lumps dropping. 'Gi ower ye bastard. Ger off is.' You try to command the trembling roof to bide its place: 'Stop up there ye – ah've telt thee stop up there.' And all the time the eyes patrol wildly the sides, the roof, the legs trembling.

Below you, too! Once anything comes down, you've got a very slight chance of getting out of there. You have to stand there and timber that up and support that. It's at times like that when you really earn your money seriously, when you think, 'What the hell am I doing here?' But you have to timber that up, and you cannot make a rush job of it, because you're going to have to pass under that place every day, and so are your mates, who come along on the next shift. If you don't you might make it worse for them, so somebody will have to get up there again, only now you have twice as much ground to support.

It's dangerous, too. You've got a scaffold, but the trouble is that you've got to hit that scaffold quicker than the rock that's fallen, and get off the scaffold before it hits it and, of course, very often that doesn't happen. You arrive there in time to get out of the way of the rock, but the rock comes immediately after and smashes the scaffold and down you all go. This is why I insist, and anybody with common sense insists, the face chain in the gate stop – because if I'm going to get crushed, I don't want to be cut into pieces as well.

But it's amazing the number of officials who insist that the face chain goes. And safety is theoretically their only job. The miners' union objected very strongly when the deputies and the overmen were put on to the incentive payments. Then they would have a vested interest in increasing the production of the face, which is often a contradiction to the rules of safety. Deputies are not supposed to be involved in production at all. That's the overman's job, to co-ordinate production.

The overman's not actually involved with supervision as such. The term 'supervisor' is directed mainly toward the deputies. They don't really have any job to do, any more; there's that many of them now. Previously, you'd have one deputy to a district. Now you've got one at each end of the face! And maybe in the middle too! You might have a deputy in the gate covering one end of the face, a deputy at the other end covering the face and deputies farther down in the gate. Timber deputies, gate deputies, all kinds of deputies. The time when deputies are valuable is when you have an experienced deputy who has worked his way through all coal face operations and is a skilled man – if you land on a face which has a great many young inexperienced lads, he can be a boon in trying to stop bad practices before they begin. Being mobile, he can keep his eye on vulnerable areas and tender-foots. The older experienced deputies will guide such young workers rather than command them.

With the inception of the incentive productive scheme, the big push for greater and greater production is rendering the harassment we experienced before even more widespread. Indeed, the deputies who are supposed to supervise safety regulations now have a direct financial interest in production, since they share in the production bonus. Where money is at stake, safety is cast to the wind and men take inordinate risks in order to maximize their earnings. Pressure from manage-ment comes down in oppressive orders and the dressings down of deputies and overmen who in turn feel obliged in some cases to pass it on. Nagging from officials rarely stops. Of course, many deputies have a more fraternal working relationship with the men, each getting on with their respective jobs unhindered by the 'supervisor-workmen' syndrome the Board would like the official to pursue.

As I mentioned earlier, the ripper insists on putting the lock-out on while he timbers the broken ground. He reasons that if by chance he does not get hit by rocks from the roof, he has a chance of living, but if he falls stunned on to the moving face chain, he risks death or at least a mutilated body to go with the broken bones. It's for this reason that you stop the chain. Also, if the chain is silent and the ripper strains his ears to the noise above him, he can utilize every ounce of his pit sense, he can do the job, secure the gate, make it safe for coal work *and* survive himself without more than bruises or cuts. This is the way we reason, the reason of responsible pitmen. But what is the response of the officials (and sadly, now with bonuses dangling in front of their noses, also the response of some workmen)? 'Get that bloody lock off. You're holding the job up, bugger that hole, let's get the coal out!' This is what we get every second the chain is stood.

The sheer non-concern of the management is really staggering. Some of the most horrific accidents have been on manriding operations. A matter of days after the Bentley Colliery paddy disaster in which seven men were killed, while the papers were still full of the horror stories, while top-level investigations were in full fling, in pits everywhere in the country the same rules were being broken continually. At Hatfield, rope paddies went in and out without lamps or conductors, breaks were wedged off by wooden chocks, pre-start warning systems were rendered inoperable. All of this within days of a major mining transport disaster seven miles away! The officials of the mine were in full knowledge of the whole situation, contrary to the big show of concern on the TV and the pretence that the Bentley accident was totally out of the blue. They faced objections by the men with the cynical, 'OK, walk!' Faced with this as the alternative to a ride, albeit a dangerous or possibly fatal one, most men will ride.

Of course, the number of accidents on transport is directly correlated to activity on the face, and there is no escaping from that fact, even though the Board in recent times have tried to deny it. They tend to defend their incentive scheme, by trying to prove that the increase in accidents since its introduction has nothing to do with the face. But even the Inspectorate of Mines points to the increase of transport accidents and relates it to the fact that the more sophisticated (and dare we add, the more hectic) operations still rely on antiquated transport systems.

There is little doubt in the minds of coalface workers, and

everyone else who can observe the day-to-day changes underground, that the introduction of the incentive scheme is
responsible for the rapid rise in deaths and accidents. Now
that bonuses can affect the wages a man draws by £60 and
£100 a week, the push for coal has become a fury. Safety
standards which once meant a nudge or a wink and a little
attention, now are totally ignored. There is such a rapid
momentum of individual toil that men are turned against
each other. 'Lazy' is the most cutting insult down a mine;
'weak' is another, and these terms are easily turned against
the safety-conscious worker, against the union militant. And
not simply abuse, for when £70 a week may be at stake, the
threat of violence – or worse, social ostracism – is not far
away, as a team of money-mad workers (previously blood
mates) refuses to work with an offending malcontent.
Principles are cast to the wind and the vilest motives of
laziness and work-shyness are attributed to those whose pit
sense goes beyond coal production. The 'dog eat dog'
attitude has come back to the darkness of underground
labour years after it was exorcised.

The Inspectorate admits that under the present incentive
scheme, there are more deaths. But the Coal Board
vehemently denies any connection, saying that the deaths and
accidents are away from the coalface. Still, it is hard to hide
the very stark changes in the underground environment,
which have changed with the way the industry is organized.

In the mad rush to get men on to the coalface after entering
the industry, the whole training schedule has been slashed to
the barest minimum. Men and lads who are strangers to the
coalface now have reduced training periods. The margin of
safety in the earlier scheme was 'taking too long'. Again, in
the choice of safety or production, the latter has won out.
The length of face training prior to 1979 (the year of the
incentive agreement) was 145 days, and the training one-to-
one. Now two of the weeks of training are at a training centre
(not on a coalface at all), and there is only one inspector to

five trainees. Much of the area of training is now subject to local interpretation which, in the bulk of cases, has rendered the term 'coalface training' almost meaningless. Supervision on developments or enlargements can be so corrupted that extraction of six inches of floor below or mud suffices as 'face training', which, of course, is in no way similar. Previously the numbers of trainees per job were tightly limited, for example, one trainee per face lip, but now with the deliberate dilution of standards, you often have one trainee in the gate and one in the pack.

In one week at my pit, there were three serious reportable accidents. In one accident, the previous branch delegate measured a complete collapse of a roof for nineteen feet with two of our colleagues underneath. Where the regulations said no more than eight foot six, they had cut a distance of twelve feet in excess of the front prop on the chock. We've no need to speculate on what caused it because that dramatic change came about with the change in how the men earn their living, risking all for the bonus.

Fortunately, the pattern of dangerous, hasty work has not continued. This has let the NCB off the hook, and allowed it to say that, in the long term, the incentive scheme has not led to a decline in safety standards. In a sense, they may be right. In so far as the scheme has proved unable to provide a decent standard of payment for the work put in, many men have just stopped trying to make the scheme work. Our old fears that the men in the easy conditions would rake in the money while those in hard, difficult, and dangerous conditions would be unable to earn anything like a fair reward has, in fact, now been proved. This has had a beneficial effect, since the men now just plod along in their old way, doing their job as correctly as they can. If they earn a bit of bonus, all well and good, but at many collieries the men have learned not to expect any fortunes.

5 PILIKING

Cruel things happen underground, and the men must steel their minds against thoughts of what might happen. All day long a man is prodded and struck by supports and rocks, working in heat and stench. Each injury sends a jolting presentiment through the mine. How does it feel to go back on to the cage the day after your partner has lost an arm? It requires a different attitude of mind, a different temperament. There is that strained quiet before going down the pit, when the miner comes in through the clean end of the pit bath. He takes one cigarette and one match with him. He strikes a match and he smokes a cigarette; that's a part of it. But there is also the other side of the quiet, the 'piliking' as they call it in Yorkshire, a funny kind of mockery. It's making fun of people and it can be very sharp. Part of the piliking process is to get somebody really annoyed, to break the tension. When this happens, they say, 'You've bitten. You've bitten the bait.' Then they stop.

The quiet before the men go down the pit and the piliking in the cage are part of the same process. As the miners acknowledge, 'A man who can't switch off his surface self and change his nature when he goes underground won't last very long.' And in fact, some don't, although there is an embarrassed silence about those who can't go back down the pit because of nerves or who take pills for a while so they can try. That failure brings no mockery, rather it seems to absorb the collective fears of all. In a war some are wounded – these battles are not about courage.

The omnipresence of danger and death infuses the community. There is the gallows humour. The pitman is out in the garden, cutting the cabbage for the Sunday dinner, his wife is in the kitchen, and he drops dead of a heart attack. The woman next door comes in and she says, 'Hey, hinny, your man's just dropped dead in the garden cutting that cabbage. What are you going to do now?' And she says, 'Well, pet, I think I'll have to open a tin of peas.'

The loitering presence of death and the permanent quality of the poverty in a pit village may be hard to distinguish. Like the gaunt, black-faced sheep which linger in the Welsh countryside where the winding gear is all that remains, undernourished, bored, scraggly kids graze in the back streets. These are the villages still alive with working pits, and if the youths look as if they have absorbed the meaning of what awaits them down the pit, it is probably only the coal soot that hangs everywhere – you dare not hang white sheets on the line – and leaves the mark of the industry on everyone and everything.

'Your blood is at the boiling point,' the men say. 'If it wasn't for the jokes, everybody would be fighting in a few minutes. The only thing that keeps you going is the humour.' And it is something distinct. It is hard to think of the pit without the comics, or to think of the village life without the ghoulish humour, or to imagine any of this and especially the constant grey, the dingy graininess of the air, without thinking that the humour's there because unemployment and listlessness are the only ready alternatives.

Humour is an intrinsic part of the pit. It just flows, it comes out of situations, totally unexpected. For example, there were four men working on the face this day, and the water broke through. And they can't get out, and the water starts to come up to their knees, so one of the men turns around and he says, 'Oh, hell, I was going to go down to the club tonight.

Now,' he says to the other one, 'we're going to be late.' So the other one says, 'No mate, it's all right. What the hell – the rescue team will come and get us out. They'll come tonight, we'll be at the club at eight, man, we'll be all right.' The water comes up to their waists, and they're all standing there, and one of them says, 'Hey, we'll be lucky to get last pint in at this rate!' He says, 'I don't think we'll make it to the dance, after all. Where are they? Where is the team? Why aren't they coming?' One of them shouts, 'Help, help!' It gets up to their necks, and one of them says, 'Do you think we'll manage a half?!' Another says, 'What'll we do, we better pray, better pray.' So, they all start saying, 'Lord, Lord, we didn't know, we didn't mean to be bad! We didn't know about these things! We didn't mean to go out with other people's wives. Lord, we didn't know! We didn't mean to keep on hitting the bottle and betting and stuff like that. Lord, we didn't know!' And just as the water came up over their ears, a voice says, 'Well, you know now, you buggers!' In an actual situation, given that that's a popular tale in the pits, I have not the slightest doubt that in a group of colliers trapped and knowing their cause to be hopeless, some wit might well chirp up, 'Ye knaa now, yi buggers!' – and the final gasps of the men would not be screams, but laughter.

We have seen in all too recent times, cases where miners have been outflanked by the growing imminence of death, with no way whatever for escape. We know now that in some cases men simply sat and awaited their impending death. What else could they do? They could panic and rip each other – or themselves – apart, but there is no evidence of such desperate gestures. The miners would sit and their tales would be told, while death crept up on all sides, and those tales which carried them through life would carry them off into the darkness with a mirth not of acceptance – but of defiance.

Miners, with the dangers they face, actually react like that. They don't suddenly turn funny or start running around crazy like a chicken with its head chopped off. Once the situation is inevitable, they do submit themselves to it. And some of them are so blasé about it! There was one particular man, who was very, very conscientious. You used to get a lot of people like this, over-conscientious. I mean, this man had sixteen stitches in his head one day, and he came back to work the next day with his head swathed in bandages and his helmet stuck on top. He was very accident-prone as well. Well, he got buried on the face – and nobody knew it. He was under this gigantic fall of rock. The men were going off the shift, and somebody said, 'Where is he? Where's Billy?' 'Last time I saw him he was working over there.' 'Well, hell's flames, has he gone out *that* way?!' He hadn't got out the other way. So they set off looking for him coming down the face. And just as they came past this pile of rock, they saw a little glimmer of light coming out of the rock. God knows how long he'd been buried there. So all of the men start tearing at this rock to break some of the rocks up. They had to get the windy picks and everything. Anyway, they pull him out, and they grab hold of him. 'Are you all right, Billy, are you all right, Billy?' And he starts digging away at the rock! The fella was just pulled out!! So they say, '*What* are you doing?!' He says, 'I left my shovel under there.' 'Don't mind your bloody shovel!!' 'Naw,' he says, 'it's a bloody good shovel!' He'd been buried for an hour and a half! They dragged him out and the first thing he did was to start working to get his shovel!

Miners are ventriloquists – South Wales miners, Northumberland miners, Durham miners, Yorkshire miners, Lancashire miners, have this tradition. There is nearly always one, in a company of maybe twenty, who's a brilliant ventriloquist. He doesn't have a dummy; he doesn't do anything like that. What he does is throw his voice out of areas. I was sitting on the face one day – I was chocking and we were sitting down this time, and from the goaf behind me, I heard this voice, 'H-e-l-p.' and I look into the gob – nobody else moved – I turned around and I say, 'Bloody hell, I thought I heard a

voice coming from there.' Then I hear, 'H-e-l-p!' a bit
louder! So I shine my light and I look into the gob and I can't
see anybody, and I'm saying, 'Hey, there's somebody
shouting in there!' They all say, 'There's nobody shouting in
there!' And then he goes, 'HELP!!!' very loud! And I say,
'Yes, there is! There's somebody shouting!!' I'm clambering
between the chocks. And they pull me back and they say,
'No, there's *nobody* there.' But they never tell you it's one of
them sitting there, doing it!

One of the best was a man called Arthur who is retired
now. He's standing in front of the shot-firer, giving him a
pinch of snuff, and the shot-firer's taking a pinch of snuff,
and from behind the shot-firer comes the shout: 'Shot-firer
wanted!' And the shot-firer turns around and says, 'I'm
coming – I'm just having a pinch of snuff!' And it's Albert,
even with his snuff-tin, that's doing it! And he keeps
shouting, 'Bloody shot-firer wanted!!' And he says, 'Hang
on a minute, will ya?!'

Another time, the telephone's ringing, and the deputy will
come running up, pick the telephone up, and Albert will
project his voice into the earpiece of the telephone. So, he's
saying all kinds of things like, 'What are you bloody doing
down there?' 'What do you mean, what am I doing down
here?' Or about the job, going: 'How much coal are you
getting?' 'Oh, just started.' 'Well, that's not good enough!'
Or, 'Tell the men they can all come off the face.' 'WHAT?!!'
This is the union secretary – tell the men to get off that face.'
And someone would stand up and say, 'I heard that!' And
the men would pretend to come off!!

The humour is what keeps you going. I can't think of the
pit without thinking of all the comics and the way the blokes
are. Just recently, I was the victim again of the Yorkshire
piliking. Being a branch official now, I was doing an in-
spection and when we finished, I was coming out with the
day-shift men. At various points along the way, the paddy
picks up men. I was on my own in a coach until I reached a

point where six very irate beltmen* got into my coach. They were determined that they no longer would work with each other: one set had had too much overtime, the other too many water notes*; one set got the easy jobs, the others were useless; one set were ugly, the other smelled; blows had been struck on numerous occasions. Bits and pieces of their water bottles and bait tins were slung out of the coach by the warring factions. All of this I was urged to write in my note-book, and by the umpteenth point of contention, with my book filled to capacity, we neared the pit bottom and I said, 'Now, look lads, this is an internal union row. Don't involve the gaffer and I'll arrange a meeting for you all as soon as we get up the pit. We'll thrash the whole thing out.' Just as the paddy stopped, one of the protagonists said, 'Oh, I wouldn't bother, Dave. It was just to keep us from dropping off, you know.' They all smiled, and there I was with a spent note-book and writer's cramp. Just so they wouldn't fall asleep!

Another thing, people wear old things for the pit, the oldest things they can find. I wouldn't take anything that had the slightest bit of use down the pit. It has to be really worked out and old before I'll put it on. One of my mates was passing by a junk shop, an old bloke, he's got to be in his fifties. He seen one of these old dinner suits with the long tails, so he went in and bought it! It was about thirty pence or something, and he put it on to come to the pit in, and he also brought with him a black stick, so he's walking along with his 'tails' blowing out behind him, and bowing to everybody like a gentleman. He was going to get himself a pair of spats to go with it, but I don't think he ever did.

Then there was this big Scottish bloke, he pulled his towel from where it had been drying in front of the fire, overnight, and his wife's nightie was there as well, and in his tiredness, he rolled them all up together. So, consequently, when he arrived at the pithead baths, and he was getting changed, he unfurls his towel to wrap it around himself, to go through the dirty end, and lo and behold, out comes the nightie. 'Oor lass's nightie,' he announced in shocked surprise. It was one of those frilly negligée things. Before he could touch it, one of these great big hairy miners grabbed it. He was naked of course, put the nightie on, and went running around the showers, chased by the Scotsman. 'Give me oor lass's nightie back!' These things really keep you laughing all the way through.

Of course, there is a lot of piliking directed at me. I must be the only vegetarian coal miner in the world, or in England, anyway. The men used to have me on a little bit about this at first. Also, I used to have very long hair. You know, I had long hair, I was a vegetarian, and I also started to do karate. So when they saw me walking up the hill, towards where the paddy was, and the wind used to be blowing my hair out, they all used to sit there and whistle the tune of 'Kung Fu' like Kane walking with his bag on the television! I used to have my bait poke* as well – and if ever they wanted any wood, and they needed to shave off a piece, they used to ask me to smash the wood.

They used to set traps for me because I am a vegetarian. Once, the men on the preceding shift hauled along this polythene bag filled with water, with goldfish in it, down the pit. But not only that – they also had put two holes in it. So, the water was coming out of the bag. So, there I was, down the pit with two apparently stranded goldfish in a polythene bag, with the water leaking out! I was swearing my head off at what these men in my shift had done, just to get at me, the indignity they treated these goldfish to! So, I'm running around everywhere, trying to patch up these holes in this big polythene bag, and fill it up with water again. I was going to go out of the pit, in fact, I was on my way out of the pit to take these goldfish out, when I found out that it was two pieces of orange peel that they'd cut into the shape of a goldfish. And for all the world, in the half light down the pit, of course, you shine it through with the water and everything, they appeared to be swimming around!

In fact it is true that a miner always feels a certain kind of sympathy for everything that shares his environment. There are relationships down the pit that'd seem very strange any place else. Like the mice, who we get used to day in and day out. That's the meaning of a cartoon done during a recent battle to stave off a new round of pit closures. We organized a massive lobby of our Executive who were meeting the Coal Board. Eric Booth, the pitman cartoonist, did a sketch showing thousands of pit mice with their suitcases packed marching off to the lobby, led by a big mouse with a placard saying, 'It's our fight too!' Now, that's solidarity.

Of course, there were strange experiences with animals down the pit at times. I was working the colliery holidays one year. This is the time, with most of the men off work for two weeks and leaving no food about, when mouse mortality underground reaches its peak through malnutrition. I had told my daughter, Emma, of their fate, and next day found a big bag of guinea pig food packed up in my bait bag with a note from Emma that this was for the poor mice. Well, throwing the odd scrap of bread crust was one thing. The men might get used to it, but a bloody great bag of assorted rodent goodies was quite another. When I got to the pit bottom, I discovered we were going up the loader gate, everyone. There was no chance to feed the chicks on the sly. So I announced to one and all that being a safety-conscious person, I would walk in the (longer and hotter) tail gate to see that all was standing well and gas wasn't accumulating. Having thus freed myself from the other men, I set off in-bye, sowing the food to the left and right like a farm hand.

I had got about one hundred yards when I became aware of a growing commotion and looked behind me. Running along the tracks, skipping along the struts on the ring sides, and bundling over the sleepers* like a steeplechase was a mass of little red eyes. I dropped the bag of food and took off like a rocket. After getting onto the face and rejoining the men, I thought no more of it until about half-way through the shift. Mice started to appear all over, climbing out of the gob, running down the chock tracks and all over the face chain. 'Eee. A'h wonder what's caused that?' the men asked each other and in feigned innocence, I too expressed total ignorance as to why the normally docile mice should be behaving in such a strange fashion!

Of course, the main animals down the pit were the galloways, the ponies. Most of the time, the men had a good relationship with their ponies. It's like the pit in general. There are disputes, there are big periods of tension, but there are also periods of comradeship. In fact, there's a constant comradeship, a deep bond. And there is affection for the galloways, the ponies. People have told me that when they used to jump on the belts at the end of the shift to ride out, their galloway used to jump on with them. They used to actually go bobbing along down there on the back of the belt. Some of the things that the ponies did you couldn't help laughing at. Like their boffins – the big water bottles they used to make them carry. The ponies didn't like the feel of them, so they used to deliberately kick each other to smash them!

A man in Yorkshire tells me that everybody agrees that his pony used to chew tobacco! When he pulled a chew of 'baccy off he used to give his pony one – he used to chew all day, and spit, you know! The pony could be just as good or as awkward as the driver made him. Though to be honest, certain of them had their own will; and some of them couldn't be bribed or bullied into it. You see, the affection for the pony didn't come until after the pony wasn't involved in direct coal work and production. When the pony was involved in getting the coal out, and the whole of the wages depended upon that pony, there wasn't much affection towards that animal if it wouldn't work properly. Same thing with a man, if he wouldn't work properly. And if a pony couldn't, wouldn't work, it was beaten. Many lads didn't like it. My father was a putter in the early days of his life. He remembers crying many a time when his pony was beaten by hewers because it wouldn't work.

I've seen ponies badly treated, as well. I know in Yorkshire, for example, ponies were only supposed to pull three tubs out, shackled together. There's a story about a pony called Shot. This Shot could count. When it heard three chains getting put on, linking the tubs together, it was all right. But if he heard a fourth chain, he wouldn't move. That meant an extra tub. If a pony was stubborn like that, something had to be done. Ordinarily, he would come belting down the line with the tubs on, the air door would be open, and he'd go straight through it. To teach him his lesson, they used to close the air door. And the pony, which expected the door to be open, would come belting down the hill with three tubs of coal behind it and be crushed against the doors. I suspect that was more common than people ever talked of, because it was money. It was time, it was money. People do funny things for money when their family's income depends on it.

6 HOLDING HANDS

I was told of an incident, at a colliery near Hatfield, where I was working. Whether it really happened or not, the miners believed it. A man was caught up on the coalface, by one of the machines. The following shift came along and knew there had been an accident, but didn't know what happened. The men were still out on the face, and weren't told. Then they were all lined up in the gate, and they were told, 'Well, lads, there's been a tragedy, a man was killed here today. Now, just go about your ordinary business. Oh, and by the way, keep your eye out, and see if you can find his head.' They hadn't found his head. And they wanted the men to *carry on cutting coal.* As if he were just so much bloody pork.

This is the kind of indignity that's perpetrated against the miners still. The whole question of injuries down the pit – this is never dealt with. Most people don't realize that when a man gets hurt down in the pit, when he gets injured down in the pit, *his money stops.* They don't even pay him to the end of the shift. A man's working down the pit, on the coalface, a rock comes and breaks his back, they carry him out of the pit, and from the time he gets out of the pit, they stop his money. By the time they get him out of the pit, he may even only have a half an hour to go to the end of the shift; they won't even pay him that.

Now, when a man gets hurt, his marras have to carry him out. They're the ones who have to tend to him. Nowadays, they'll pay us our money for that, although sometimes they don't, if you carry him out of the pit and you go off the face. If you insist on staying with your mate when he's hurt, then definitely you won't get paid, if the deputies don't want you to. If you say, 'Look, *I'm* taking him out, he's *my* mate,' then you definitely won't get it. Because what they're trying to do is to get a couple of lads to carry him out; if he's not too badly injured, to walk out with him, if the man's got his finger cut off, or something like that, and he's capable of walking. They'll maybe send two lads out the gate with him, rather than take his marras who he feels more secure with and who are more able to deal with him, if he passes out or anything goes wrong. If you insist on going, you won't get your money, no matter what happens.

But usually, the man's mates just automatically go with him, they get him on the stretcher and they get him out of the pit. Carrying him out can be hair-raising. You have to go through all these conditions with a man on a stretcher. Some of these men are bloody big men! And me a little fella, your arm feels pulled out of its socket at the end of the day. Your mate on the stretcher, maybe in agony, maybe vomiting, maybe covered in blood. And there you are. It can take an hour, an hour and a half, to get out of the pit. If you're lucky, if you're lucky.

Take the lad with minor cuts and bruises. Suppose he rides out of the pit, loses his time, goes to the doctors, and signs up on the sick. He loses his wages, obviously, but as well as this, it is unlikely the doctor will sign the man off for more than six days. He has three waiting days in which time he gets paid nowt, then he'll claim three days social security or sick money. With luck, it will take him just under a month to catch up with his finances. With all these penalties, the miner usually chooses instead to work on, black and bloody. It is during this process that we often acquire 'the miners' badge of slavery', the indelible blue scars on our bodies. These come about as a result of coal dust entering the small and medium wounds. Coal, being antiseptic, heals in the wounds without leaving a scab. Even the coal doesn't scab in the pits!

A good case was Johnny Davo, who, a few months back, was nearly killed in a fall at Hatfield Colliery. His head was split from end to end, and as he staggered out of the darkness, he presented a terrifying scene. He required thirty stitches in his head. After two weeks on the sick, the doctor sent him back to work, back on the face lip, despite his protestations that he was neither physically nor psychologically ready. So if you know you're gonna get two weeks off for thirty stitches in your head, it's for sure you'll get nowt for a cut arm; so the men work on. Of course, when they do, none of them make the statistics as accidents.

When there are serious accidents, the whole thing's a matter of chance. You tell them a man's been injured, and you want a paddy to come in. The exchange* man has then to contact the paddy driver. The paddy driver doesn't have any short-wave radio and surface radio waves don't naturally come down the pit, i.e. if you had a transistor radio underground you couldn't pick anything up on it as those surface radio waves wouldn't travel down there on their own. You'd need a bloody big aerial up the shaft to pick them up. Often there's no locos left at pit bottom. They may all be engaged somewhere else. So, the only check, then, is the phone-in system, which the man does when he's finished his operation. People don't get hurt at regular times – they can get hurt at any time. So there's a chance on that. There's no permanent paddy left there for people who get injured, and even if there was – how many paddies can you leave there against the number of people who get injured? So, you have to wait on that. Also, you have to hope he doesn't get off the road, coming in-bye, which can happen. You cannot go at breakneck speed. There can be an incident in the roadway to cause difficulties. Sometimes you have all these things happening together; that's usually the way of things.[2]

You carry him out . . . but there's nothing so devastating as going into a shift and seeing men coming out with a man on a stretcher. That's really inspiring! Yet, even in this kind of atmosphere, people make light of it. One of the lads we thought was dying this day, we had him on a stretcher. We could only walk every ten yards. We had to keep putting him down. For a start, he didn't like to be carried the way we were carrying him – we had to turn him around the opposite direction, so his feet were in the direction he was going, rather than his head. This was really upsetting him for some reason. Often, of course, when you're feeling bad looking straight up at the roof, that doesn't help much, either. And don't forget, you're not walking along a straight path – you're walking along a fell walk, where the roads are up and down and everything, and you're being bumped around. There's still water dripping on you, and all the rest of it. We actually thought he was dying a few times – we'd give him the 'kiss of life' and a heart massage and everything. We had to send for a nurse to come down.

The nurse comes down as a prelude to the doctor. She's called in if it's something very serious, or if it's a case of somebody going to die. Immediately, she follows the miner's crack – the miner's way of making light of it – she goes straight up to him, and says, 'You know, we'll have to stop meeting like this!' And he's a big, fat fellow, and she says to him, 'Is that all *you* under that?' (She's lifting the blanket.) 'How many of you are in there?!'

If you're working with a man who has an injury, a serious injury, like he has his hand cut off or something, then you have to dress it. The deputy might come down the face eventually, but it's a long way to come, and even then, you've got as much idea as he has usually, what to do. You have to dress it, and you have to keep the man happy. You take the man out, you go home, and you get up the next morning, and you think, 'Good God, I don't want to go. I don't want to go back there.' They talk about absenteeism. So you have a day off, something like that. It might hit you two days afterwards; you think, 'Jesus Christ, that man's got no

hand.' Or, 'He's got no arm.' Or, 'He'll never walk again.'
These things affect you in many, many ways. In ways you
might think are strange. But they aren't, not really. To
survive, you have to make light of the whole thing, you have
to.

Sometimes the lads can be quite cruel. One guy they
weren't too fond of got a leg injury and there was a fair bit of
blood. After I had got the lad dressed up with bandages, I
assured the crew that although it looked bad, it wasn't too
serious. The lad of course felt like his unfortunate life was
coming to an end and groaned all the way out-bye on the
stretcher. 'Shut up, you moaning, fucking pig!' was the kind
of reassurance he got on his journey out-bye. Worse than
this, as we tried to lift him over a belt, one of the lads lost
his grip on the handle and the poor bloke rolled off the
stretcher on to the ground. Well, of course, this was cause for
great mirth, and while he rolled around like John Wayne in
the final shoot-out, the others were hysterically laughing.
After much tomfooling, putting him on the stretcher face
down and inside out, we finally got the poor bugger to the
surface.

On another occasion, one guy thought to be fairly seriously
hurt had been strapped to a stretcher for the journey out-bye.
This journey involved a hike with the stretcher up a 1-in-3
drift. Half-way up the hill they see a runaway tub belting
down towards them. Well, they dropped the stretcher and
ran. After the tub had careened past them and crashed, they
rather shamefacedly crept back to see what had become of
their mate. No sign of him anywhere. After a search they
found him, bolt upright, with the stretcher still strapped to
him, standing up in a refuge hole as white as a ghost even
through his pit black.

And, of course, because there are so many characters down
the pit, and many deaths down the pit, the coming together
of the pit character and the fact that there are deaths
produces also a whole legend of ghost stories and ghost

phenomena. Now, some of these have to be taken very seriously into consideration. At Hatfield, there's one particular district which was notoriously haunted. I mean, very rational people came screaming out of the pit. One lad in particular, I remember, asked never to go down again. He kept on seeing this character sitting in the corner! The union man had to go down with the men and sit there – he wouldn't go down on his own. There had to be one Coal Board man and one union man, sitting there, holding hands in the dark, waiting for the ghosts to appear! And when they said, 'We never seen anything,' the men knew it was because they had their eyes shut!

At Wardley, there was a man used to have the job watching the sump, which filled up with water. When the water got to a particular level, he had to turn the pump on. That was the only job he had. It was a big heavy pump; Wardley was a water-logged pit. That man, he used to go down the pit on a weekend, and he used to be down there on his own. So he used to take his boots off, and his socks off, and lie with his feet over the end of this sump, and go to sleep. And when the water touched his feet, it woke him up, and he could put the pump on. It was like an alarm system.

But one day, early in the morning, he woke up and felt himself being dragged! And there, towering above him, was this character who promptly grabbed ahold of him and pulled him into the sump. He had water up to his neck. A ghost! And the man's screaming at the pit bottom, and yelling. At Wardley, you could shout up the shaft and you'd be heard – you didn't bother with a telephone. He was just screaming, 'Let me out, let me out!!' And people had to go down and look at them things. He was serious.

Another haunting at Hatfield was the image of a pair of boots walking along, all by themselves. They would suddenly appear in front of the line of men walking in-bye. Of course, some of the lads turned even this into humour. I was walking along and a bloke says to his mate, 'What wad thy dee if tha seen the boots appear?' To which he replied, 'A'd say, had-up marra, let's swap them buets, these buggers is killing me.'

7 PNEUMOCONIOSIS

In the villages, one sees men who haven't got the wind to walk the hundred yards to reach a bus stop. You see them at the foot of a hill, gesturing to passing cars, waiting for a ride to level ground. Then they walk the remaining distance home, using their handkerchiefs and looking away. They characterize the mining village – the old men, spitting and gasping for breath. Many of them are still working at the pit. Perhaps they don't like to admit their problem to their mates or to themselves. They need further disability, in any case, to receive compensation. Yet, they are nearly done. You see them walking up the steps – it's always peculiar: you have to walk up the steps to go down the pit, up to approach the cage. Miners talk about that: 'We always find that funny, to walk up to go down.' You find them walking up the steps 'til they get half-way, and they'll pretend to be admiring the view, looking at the countryside. And they're not, actually, they're fighting for a breath. Climbing up the hundred or so steps, they seem to be puffing harder each step, until the agony of the last three or four cannot be concealed. Then they will say, 'If only they would get rid of them last three stairs, I'd have no bother getting up here.'

And you see them coming down the pit lane, having to hang on to the fence. As their conditions worsen, in time even small exertion becomes impossible. Coughing brings up blood, their weight drops off. Lips and nails darken, and then the rims of their eyes. The skin colour, if this is possible, seems blue. Finally, the heart is affected, ankles and liver swell. The miners will talk of this amidst great silences. What can an outsider know? Death, they say, follows this stage after a year or so.

There are a number of ways to speak about the hazards of work in mining, and all have been tried. In February 1974, for example, in an unavailing effort to avert the strike which was to bring down the Heath government, the miners and the NCB were required to place evidence before the Pay Board. The NUM presented a brief which explained the nature of their job and underscored the difficult working conditions. To give their argument more substance and objectivity, they included an unsolicited statement issued by Dr. Hugh Faulkner, the Medical Secretary of the Medical Practitioners' Union. He chose mainly the statistical tack:

> Our Doctors serving the coalfield communities have urged us to make public certain disturbing statistics about the miners at work and in ill health. Those who deny the miners a special case have to be reminded that in terms of death and permanent disability from accidents, occupational chest disease, sickness and stress, they carry greater all-round risk than any other group in our community – except perhaps deep sea fishermen. The mortality figures given by the Registrar General speak for themselves. The last available figures showed a large significant improvement *in all mortality* for miners. For this we have to give due credit to those in the trade unions, the Coal Board and its medical and technical services who have long battled to improve conditions in the mines, but the comparison between miners – in particular coalface workers – and other groups of workers still shows a startling relationship.
>
> Comparing the standardised mortality ratio for miners with all other occupations, miners are 15 times more likely to die from occupational lung disease – and nearly

4 times more likely to die from Pneumoconiosis complicated by Tuberculosis, 5 times more likely to die from accidents, and nearly 3 times more likely to die from Bronchitis. While technical advances have improved working conditions at the coalface, mechanisation and increased production have increased stress. Improved measures to control dust have barely kept pace with increased dust from high-speed cutting machines. Though conditions have improved over the years, the figures for miners and their wives still show the main causes of death as Tuberculosis, Rheumatic Heart Disease, Endocarditis and Bronchitis which are not diseases normally associated with affluence. From the Registrar General's figures, we conclude as Doctors that the miners are engaged in a highly dangerous occupation in the vital service of the community – in many respects the most dangerous. As Doctors we see the miners as a very special case demanding the understanding of their fellow citizens. We appeal, on the basis of these figures and the evidence from our own members, to the whole community to urge the Government to give the miners the special consideration their case demands.

The union supplemented this statement with what they characterized as 'hard statistics':

Since nationalization, (they reported) 7,718 men have been killed in the industry – 80 of them in 1973 – between 1951 and 1971 more than 17,000 died from Pneumoconiosis. In spite of the high safety standards achieved by NCB/NUM initiatives, the damage to health is considerable. In 1973 some 626 new cases were diagnosed as suffering from Pneumoconiosis. In the same year 569 men were seriously injured.

The number of reported accidents each year (excluding

pneumoconiosis) is equivalent to one-third the number of miners.

It is as if the union officials' confession to the faith of modernity forces an unwitting accommodation in the way they talk of death, through numbers. Sharing leadership in a technological industry means accepting the logic of figures. The data are so striking, that the distinctive character of the personal tragedy is nearly lost. Not everyone is constrained by such an obligation, however. Take a statement of a rather different kind, from a group of widows of US miners, whose statement speaks for British miners' widows too. (Miners' wives in Britain, incidentally, suffer a 29 per cent excess standard mortality rate.)

We are widows of coal miners who gave their lives to the coal mines in Eastern Kentucky. The mountains are full of women like us. From the time they were just little boys, our men worked long and suffering hours in the mines. They were paid so little they would have to work double shifts so our families could survive. We watched as their health suffered from the cold and damp, as their bodies became broken from accidents, as the dust got their lungs. Next to the miners themselves, we know best how men suffered and we have suffered greatly ourselves. Because the men had to work so hard in the mines, our work at home was doubled. We had to raise the children alone; we had to do all the gardening, take care of the house and the animals. We had to fetch the kindling, carry the coal and water. We got our families through the hard times on nothing more than pure 'stick-to-it-ness.' We were at home when the men had to be picked out of the mines on account of 'bad air' but when they started suffering so bad from the dust we had to nurse them and care for them with ourselves sick and tired. For years we would lie helpless at night listening to our men smothering and coughing, never knowing when we would wake to find them dead. Before and after the men died, we had to work at jobs ourselves – in factories and homes wherever we could. All of this we did for the sake of our families. The country has prospered while we have suffered and continue to suffer. We have lost our husbands and are lonely now. Most of us are sick from the years of hard work. We can't work anymore.

Pneumoconiosis – black lung disease – is, of course, the miners' unique tragedy. It has to be appreciated that the disease has no immediate impact. It is not an accident. It takes place over a number of years in tens of thousands of little accidents inside the lung, where the dust particles cut the fine membrane, causing scars when healed. Detection and evaluation of the extent of damage is difficult and fraught, given the far-ranging implications of medical testimony. The financial weight of compensation is at stake in each case, as is the reputation of the industry for potential new entrants and for a population at large which is routinely called on to choose sides during industrial disputes. A doctor who endorses a claim for black lung compensation defaces the imagery of the 'new' coal industry. And if the historical truth of mining is affirmed by scientists in white coats, then who can blame those in pit helmets and cloth caps for waging their century-old battles once again?

So the tragedy of the dust, and its significance, makes for complicated procedures. For the purposes of evaluation and compensation, the disease is divided into simple pneumoconiosis itself, defined by three stages. If a man is diagnosed as having stage two, or stage three, or PMF, a government compensation scheme automatically comes into operation. He is medically assessed and his benefit determined by how many black spots can be counted on his lungs. These are measured in blocks of 10 per cent up to 100 per cent. Compensation depends on X-ray evidence.

The category one sufferer is wholly neglected. These men are told, 'You don't have enough dust.' Of course, they can oblige objective medical requirements by developing the next stage. Tragically, once stage two is reached, men have a substantial possibility of acquiring PMF, even if they come permanently out of the pit.

A man applying for certification of pneumoconiosis goes before a medical panel, comprised of functionaries working with formal guidelines. The principle seems to be criteria of compensation rather than effect of disease. This leads to some jesuitical distinctions. A recent NCB medical service report explained, 'It must be remembered that category one pneumoconiosis is the earliest radiological sign of dust retention in the lungs. It is not generally regarded by medical opinion as being a disease process.' Actually, medical opinion is substantially divided on this issue, but even that fact misses the point of the manifest suffering at every stage and the indignity of the process. One can see a miner collapsing of breathlessness going back down the pit because he received no compensation, since the shadows on the X-ray plates – not the suffering – define the disability.

The evaluation is complicated further by the presence of other respiratory diseases, like chronic bronchitis and emphysema, for which miners receive no compensation. To miners, it seems clear that these ailments are either caused or aggravated by the pit, and they are convinced that when a miner has dust and chronic bronchitis, the Medical Board will plump for the non-compensatible ailment, diagnosing only bronchitis. Who can doubt that the presence of bronchitis or emphysema will hasten the demise of the sufferer from dust? Yet, the assessment of pneumoconiosis for purposes of compensation is made by subtracting the value of the accompanying conditions from the overall degree of disability.

The miners' conviction that they are cheated of appropriate compensation for pneumoconiosis is exacerbated by their general mistrust of medical officials. It is hardly surprising that miners and doctors greet each other – both in disputes about dust and in accident cases – with mutual suspicion. The miner feels that he is being perused by a person who knows little of his life and labour and cares less, the doctor that he is being used to get time off for a malingerer or malcontent. Both professional reputations are at stake, but only one is recognized. A telephone call from the pit manager to the local GP is not uncommon; for example, warning the doctor to be wary of an inrush of miners wanting notes off work since a local strike is brewing. Actually, the opposite is more likely, explained one Yorkshire GP with long service in a mining practice. 'Men who were on the sick before a strike occurs actually asked to be signed off during the duration of the strike,' he observed. 'They don't wish to receive financial benefit while their comrades are involved in a dispute and receiving nothing.' But the mistrust is pervasive and may lead to dangerous situations. Miners with back injuries, for example, are loath to go to doctors with their complaints, even though back ailments are very common, since doctors cannot readily verify them and suspect that a con is being tried. The aura of the middle-class professional facing the dull, ill-clad working man deepens as the doctor makes little effort to disguise his suspicions, and shows little sympathy for the patient's complaint. The miner remains muted by class deference and the doctor's trappings of authority, outraged by the 'Oh, yes, you're here again' response. This leads to a strange situation. The miner routinely revives old verifiable injuries from which he is not at the moment suffering, rather than admit the genuine back injury which plagues him.

You've got the dust that's coming from the boring operations; and from the firing operations. You've got all the coal dust coming from the coal-cutting machines, and all that's coming down the face; you're breathing that at the tail-gate end, all the time. Plus the heat, because the air's hot by the time it gets there. As well as all this, there's the excessive shovelling, and you've got all of this bad air as well.

Now, with the new operation, with the boom ripper,* this replaces the boring and firing, and it actually cuts out the rock. The trouble with this machine is it creates an amazing amount of dust. If you can imagine, the boom ripper is turning a cutting disc, and it creates a tremendous amount of dust — it's cutting into solid rock! And, of course, all the dust comes down the tail gate, so the men who are all working in the gate back-bye* are getting all of this dust. At times, the dust is so incredible you just cannot see a hand in front of your eyes — it's literally true. You cannot see your hand.

Take the example of the airstream helmet*. Nothing helps. By March of 1982, b.12s (that's the name of a coalface unit in the Barnsley seam at Hatfield) had become a black nightmare for the men working on that unit. A thick, impenetrable deluge of coal dust speeds down the bank from the cutting machine and the heavy duty chocks which crush the coal roof to powder. The first test of the airstream helmet has proved abortive, here, as the men say it is impossible to wear the helmet and work. The men have put up with the conditions for nearly two years to make the unit work and have now reached the parting of the ways with the gaffers. They insist the stall* be stopped. The men are easily recognizable in the pit bottom at the end of their shifts since they are not simply black like other faceworkers but are caked in thick coal dust from head to feet, inches thick. On the surface, they are easily picked out, too, since their eyes resemble shattered glass, they are so bloodshot and raw. The airstream helmet is not something to be written off, however, and in some areas of pit work and operation, it is a godsend. It is not applicable to many coalface operations, though, particularly in cramped places or where vigorous body movement is required.

It's the same with masks. Dust masks can be worn, if you're not doing anything. So what's the point of wearing them? If we're not going to do anything, we might just as well go out of the gate. But if you're working, shovelling, first of all you're breathing at a faster rate because you're exerting yourself. The ability to be able to take in enough air is cut down by the fact that it's hot already, and it's full of dust. If you put a dust mask on top of all that, you drown in your own saliva, for a start, and you literally cannot breathe *and* work. They only give you the mask so you'll work. If you cannot work with them on, you might as well leave and sit where there's clean air! In any case, the really harmful dust penetrates the dust mask. They cannot make the gauze close enough to stop the dust that does you the damage. It's a con, it's a con. Just now we have introduced a soft fabric disposable dust mask worn for one shift then thrown away. This is having good reports from face workers at Hatfield, but it is too early to assess how effective it is in preventing penetration.

Now, they're supposed to use water on this type of cutting machine. But often people do not use water on the machine, because after they're finished cutting, the men have to shovel the stone out. They have to work there. First of all, it makes the bottom all wet and slippy when they're transporting materials. It means the workers are going to work in 'self-induced' water. Also, if it wets the top, it cracks the stone and can bring the roof down on top of you. So people will take the dust instead of water. Either that or – with the bonus schemes – it costs you too much money. In drivages, where the Dosco Roadheader creeps forward on big tank tracks, the men tend not to use the sprays, because it often results in the tracks getting bogged down and the men being prevented from earning money on bonus while they dig the machine out with shovels.

Other occupations have their dangers and inconveniences, but we know of none other in which there is such a combination of danger, health hazard, discomfort in working conditions, social inconveniences and community isolation.

Thus concluded the Wilberforce Court of Inquiry which had been created to arbitrate the claims of the NCB and NUM during the 1972 miners' strike. One can consider any of these points, in turn, and reach some informed judgment about the miners' condition. Each has a peculiar meaning, which the words themselves do not adequately convey: 'social inconvenience', for example. Although there are numerous variations, typically the miner works on a three-shift continuous cycle, passing on the way in-bye his mates from the previous shift returning out-bye. The first shift starts at six in the morning, which is the time the miner is due *at the pit bottom*. For this day shift, they get up at, say, half-past four in the morning and return to their houses at two in the afternoon. On the afternoon shift, they leave the house at eleven in the morning, and get back at half past eight at night, and on the night shift, nine o'clock leaving the house and getting back at maybe six in the morning. There is no shift which approximates a normal working day and allows for a reasonable pattern of home life.

In addition, the miners are getting paid for only seven and a quarter hours, which is the official time. The estimates above include the time the man leaves the house, and the time it takes to walk what may be only a few hundred yards to the colliery (although this pattern of NCB housing adjacent to the pit has eroded in recent years). All the time for getting the identity checks – which seems very like clocking in at the factory – changing into the pit clothes, getting the lamp, walking to the cage and travelling down the pit: all this is excluded from the 'shift time'. Likewise at the end of the shift: the men have to get on to the cage, come out of the pit and put back the lamps. They get stripped, bathed, and ready to leave the colliery. None of this is paid time. There is no explanation for this pattern of three anti-social shift times, nor for all the necessary labour in preparation for going down or going home again which the Coal Board gets free. 'The Coal Board', claim the men, 'say it's our dirt, it belongs to us, so we get washed off on our own time. We always say if it was a gold mine, it wouldn't be our dirt, they'd wash us and wring us dry and hang us up. They would also get the dust out of us. . . .' Perhaps it makes no difference, but the pitmen are particularly embittered by the fact that the thirteen men who were killed in the cage disaster at Markham Main were not even getting paid at the time.

The day shift is the worst. You get up at half-past four in the morning. That's really an incredible thing; when that alarm clock goes off – you could sail it through the window, jump up and down on it! And it's a very secluded world at that time in the morning. You get up, the place is dead quiet, you're left alone with your thoughts. And you cut the time as near as possible so you have as much time in bed as you possibly can, which normally means about four and a half hours. I used to have difficulty sleeping, because I used to think, 'I have to get up soon.' It used to keep me awake. Of course, it wasn't just a matter of getting up, but what you were going to have to do – going down the pit and everything.

It's really very, very difficult to come every day of the week – you're really exhausted by Wednesday. That's why men have days off work, on the day shift in particular. When you're out in the canteen, everybody is sitting there sort of baggy-eyed. And then you have to go down the pit. It's actually that you are going down the pit, the heat and atmosphere that gets you there. You could swoon, the way you feel, your heart's going like the clappers. You really feel ill. You're supposed to do a full shift's work, a hard shift's work as well, under these conditions. You have to remember how many people actually get injured during that time on a day shift, because you are less aware, your wits are less about you. Some men do get used to that shift, but not faceworkers so much. Older men who are working back-bye get used to it. There may be one or two tasks and they will actually have to wait for a while before they can get started in the morning, and knowing that, helps them. When you go down the pit as a faceworker, you've actually got to get to the face or walk to the face, after you get off the paddy, when you're feeling tired. And you're stumbling all over the place, and everything's an effort – everything weighs twice as much on the day shift. There were these mechanical courses for the faceworkers, and everybody was trying to take the mickey out of the instructor. Somebody said, 'Hey, which weighs the heaviest? A prop pumped up, or a prop lying down?' And he answered, 'Them buggers on the day shift.' It's true. They do feel heavier on the day shift.

And if a man does manage to finish on the face, he's finished, shifting the coal in the stable, advanced all the props and he has a half-hour to spare. He comes out at the end of the shift, he comes in the gate, he gets his clothes on and he sits down by the stage loader* to wait until lowse. He doesn't go out early, he waits for finishing time. Then maybe he'll fall asleep. If the undermanager comes around and catches him, he could fine that man for more than a shift's pay. The man's done all that work for nothing if he falls asleep because of the time in the morning and the work. And it's not a question of lying down or anything. The man is sitting in an upright position. Or even squatting. People get so tired. I remember one trainee who actually fell asleep while we were putting a girder up, and his head went over the girder, when we were pumping the prop up. And he just managed to pull his head out of the helmet, and his helmet got squished! He actually fell asleep with a girder on his shoulder while we were putting the girder up on the roof. It could have crushed his head.

8 ANTHRACASAURUS

Once they actually discovered a giant fossil at Usworth Colliery. The men were working away at the lip, hundreds of feet underground, and suddenly a rock fell away, and there was this gigantic, prehistoric animal's head and its body tailing away from the face. It was an enormous thing, an anthracasaurus. The miners thought it was wonderful. They were awed. But the officials weren't too impressed, and they said to bore some holes in it and get it blasted down, and cut down, and get on with the work. The men refused. They wanted to keep this thing. Some of it had fallen away, anyway, just of its own, by gravity. One part of it fell out and smashed. But the miners did their best to preserve the whole thing. Eventually, the whole job was held up, everything was held up! A guy from Newcastle University came along and identified it as an anthracasaurus, one of two found in the world. It was fascinating to us, that this beast had actually walked around here in Usworth and Wardley, and here it was sharing our underground environment with us, even though it was dead.

This is the kind of thing you have fantasies about, to find a *live* prehistoric creature in the old collieries of Durham, Northumberland and parts of Scotland, particularly in the shallow pits. The men will be working toward the coalface and suddenly the wall will cave in. And there, in front of their eyes, there's a district that's been worked by monks, maybe hundreds and hundreds of years earlier. There are bits of their shoes and their shovels, and trams and equipment. That's happened many times. The miners look at it in awe and fascination, at their brothers from hundreds of years before. I've always had that kind of awe, from my first days down the pit.

Most collieries are not the dark satanic mills one associates with crude industrialism. To the observer, they may register a vague foreboding of what goes on out of sight, but there is a lovely irony to the placement of collieries. They emerge from the most beautiful country surroundings, unexpectedly coming into view amidst rolling hills and gloriously idle sheep. It takes an act of will to focus on the images below, when above one is embraced by a calm, natural beauty. The further from a town, the more the men seem to sense this disjunction in their lives. Way atop a hill in western Durham, as the physiognomy of the environs began to approximate the Lake District nearby, a miner – who called himself, if this can be believed, 'Hughie the Hewer' – explained why he retired to such a remote region, with a single pub in the valley the only trace of civilization. 'I've been so close to hell,' he said, 'and I don't know where I'm going next. I thought I would get as close to heaven now as I could.'

There is also a literary imagination which augments this casual naturalism of some miners. At Teversal Colliery in Nottingham, a union official asked – as if this needed to be settled before any other business could be conducted – 'Do you know what pit this is, lad?' Receiving a non-commital reply, he explained that Teversal is the 'Tevers Hall' in D. H. Lawrence's *Lady Chatterley's Lover*. Intrigued by my ignorance of a history they held very dear, the colliers proudly explained that Lawrence, a 'Nottinghamshire man', had grown up nearby. Their feelings of kinship with him transported them for the moment from the immediate fears of the shift about to start, and NCB threats of closure, and they began to discuss – and dispute – the particulars of the novel.

Each had a theory about where the gamekeeper's cottage had been, and as they pointed one could very nearly see where Clifford Chatterley's wheelchair had stuck in the mud, humiliating the impotent Lord (a colliery owner, after all) in front of Lady Chatterley's lover (a former pitman). The book is for these men an artifact which confirms their history, and they live consciously in the reflected glory of D. H. Lawrence, their countryman.

Many people in the industry have absorbed what is written about them into their culture. An older miner, remembering the high hopes about nationalization, will quote from *Close the Coalhouse Door*, a play by Alan Plater based on stories by Sid Chaplin, a kind of contemporary Mark Twain of the North:

'When it's ours, Jackie boy, when it's ours,
There'll be changes, bonny lad, when it's ours.
When us colliers take control,
No more twelve-inch seams of coal,
No more means-test, no more dole,
When it's ours, all ours.'

The play was performed at the Newcastle Playhouse and immediately became a local classic – dignifying the miners' work and affirming by the authority of the written text and the seriousness of the setting, in the theatre where Shakespeare and the opera are performed, the disappointment the pitmen had felt, their sense that all too little had changed with the change in ownership.

In fact, in some collieries almost nothing *has* changed. Rainton Adventure is a drift mine, recently closed, in Durham. A miner out of Orwell's *Road to Wigan Pier* would have found nothing in the least unfamiliar there – cramped, uneven roof along an unpredictable downward incline for a mile and a half, now high enough to stand, now forcing one almost on to the knees, wires for pulling the tubs of coal to the surface, slick and rapidly moving underfoot. It is maddening, not knowing when or indeed whether it will be possible to stand up again. The easy gait of the miners, who are able to traverse the terrain with so little apparent difficulty, is somehow troubling. This was an ancient mine, with no cage, no machinery beyond small pneumatic picks and, to its last shift, tubs were pulled by ponies until they connected with a pulley system half-way to the surface. Geordie miners know Orwell's book and are publicists for it. A book like *Wigan Pier* matters to them: it confirms their history against the propaganda of modernism, and they are proud of it. But they are also wary, for Orwell was an outsider (unlike Lawrence), a southerner, and they think he got some things wrong. This embarrasses them – and makes them angry.

There is a closeness and an affinity in the community of miners, which has an exclusionary cast to it. Miners are a suspicious people. They are suspicious of outsiders; they are suspicious of anything out of the ordinary, and this is an important part of the village life. They always wonder, 'What's this man after?' 'What's he trying to do here?' Nothing is ever straightforward to a pitman; there's always the question, 'What's in it for you?' Thus, angry defensiveness, always mixed with admiration, directed at Orwell: both express the insularity of the village.

'He's right that walking in-bye is like climbing a smallish mountain at the beginning and end of the shift, and he's correct in his observation that old colliers when black and stripped look young,' confirmed a Geordie pitman, who then lost his temper with Orwell. 'But the naïveté of the man is at times quite excruciating.' He pulled out a worn copy and repeated Orwell's classic account of the colliers' bathing before there were pithead facilities:

Probably a large majority of miners are completely black from the waist down for at least six days a week. It is almost impossible for them to wash all over in their own homes. Every drop of water has got to be heated up, and in a tiny living room which contains, apart from the kitchen range and a quantity of furniture, a wife, some children and probably a dog, there is simply not room to have a proper bath. Even with a basin one is bound to splash the furniture.

This really set him off. 'Yes, even in a miner's home in the 30s or 40s, the furniture, such as it was, didn't stand splashing. But what did Orwell think the filthy and often bloody collier did when he came home – did he lay on this non-splashable furniture? Did he sit on it – or did he hang in a cupboard somewhere where he couldn't dirty anything? Perhaps he lay out in the back garden.'

This is the other side of the apparent open congeniality of the miner in his village. He knows what he knows and others do not. Even others within the village are scarcely above suspicion. There is a crude motto of the Yorkshire pitman: 'Fuck all them, there's just thee and me; fuck thee, that's me.' Or, as they put it more politely, 'You vote for the Labour Party and we'll look after ourselves.' Theirs is a tight brotherhood and a secret life which denies admission to all outsiders, Orwell included.

My Dad was always on low money in the latter part of his life. He'd got a double fracture of the skull and toes broken, his knees knackered* and things like that. Then his hip went. He had to have a big operation on his hip because he couldn't bend. It just didn't move. He was on sticks for a long time. Pathetic. So there just wasn't anything left over at the end of the week. Me Mam had triple bingo once a week, and maybe they would go to the club once and me Dad would have a shandy and that was all. That's all there ever was. But we always ate well; we always had food. That was the one thing that was a big preoccupation – that with all the struggles we'd been through, that we were fed.

The Methodist element and temperance was much stronger in those days. Me Dad said to me one time, 'You'd never surrender your mind to the bosses?' I said, 'No.' 'Nor to the state?' 'No,' I replied. He said, 'But you'd surrender it to a bottle!' It was a big part of the tradition. You didn't surrender youself to one of the temptations that capitalism had offered the working class. Beer.

I don't think I ever understood me Father until I went to work in the pit. Then for the first time I realized that he was a workman. I'd been preaching about the workers since I was fifteen, but I never thought of me Dad as one. That had never come to me in terms of my own family life. Then it occurred to me that *he* was one of the people I was talking about. Somehow, it was a distant thing, until I went down the pit, and I'd seen him black down the pit. Then I thought, 'My God, all these years he's struggled down here as a pit-man.'

And I've often wondered how my own child feels, and how she felt about me at the pit. At first, when she was three or four, she didn't understand. She'd say, 'Why are you going down the hole?' She knew it was a hole, she called it a hole, which is like a pit. She'd say, 'Why do you have to go down the hole?' Because she could see me visibly distressed like on an afternoon shift, when I went, and I used to come home from work knackered. Her favourite greeting every day, every single day, 'Did you get hurt today?' And I used to show her the bumps on my elbows, or the new marks on my back, 'cause I would be getting shaved, and she'd come in and see a new injury, that you get every day – all the little knocks and things that you get. Or, if you pick her up and she sees that your hands are all scratched or marked. You know what used to be my greeting every day: 'Why do you go down the hole, Daddy?' And I'd say, 'Well, in order to get food and clothes and toys.' She actually thought that I dug these things out of the ground; that all these things were down there, somewhere! There is always a pile of coal in the back garden. So it was just like playing with sand at the seashore. My little girl plays in the coal heap right now with a shovel and spade, and lugs it into a truck and takes it up and down the garden. She's got a very black teddy bear, which also helps her occasionally.

Of course, she's always known something about coal and its connection with my work. You could tell when she watched the railway wagons with the coal going through. She thought me, and a half-dozen other people, got it out, all of it! For years she didn't realize that work isn't just work at the pit. If anybody said, 'I'm going to work,' she immediately presumed that meant going to the pit. I went on the bus, and I said the driver was at work, and she said, 'Well, he's not down in the pit!' She had thought in her way that that was work, that was the only work in the world, was down in the pit. I wonder how it contrasts to the children of professional people, for example, what they think work is – they think work is going at eight o'clock in the morning?! Coming back, having your dinner, and things like that. The only consolation about being a woman in this day and age is that she won't have to go down the pit, but can do some other work. Of course, I'm aware of the women miners in the USA and the breakthrough where they won the dubious right to equal exploitation below ground. My own feeling is that we should struggle to isolate the numbers of people below ground to fewer and fewer, by lowering the age of retirement and raising the age of entry. I do not feel it is an industry women should be fighting to get into, nor anyone else, really.

When I first went down the pit, God, it was terrible. Me Dad violently objected to it. He said he'd rather I worked on a building site making tea than go down the pit – anything! Well, I had tried other work, and had either been sacked, or one thing or another; and you know you get a reputation, even at that age, you know, as a trouble-causer. So coming back and working in the pit was the only solution.

I know people who left the pits six, seven times. Every time a new factory opened, or they're taking new men on, off they go and that's it – they make a bonfire with their pit clothes. They have a big ceremony. And they throw their helmet down the shaft. Or they kick it all the way up the pit lane. And that's it – they're finished, they're free, they're away at the factory. Then six months on, nine months on, the factory closes; or there's redundancies, and that's it. The dole sends them back to the pit. When you're on the dole, you can't say, 'I won't go to the pit any more.' You're registered as a pitman – that was one of your jobs – and you have to take it, otherwise you don't get any dole money.

It's hard to leave, even for the young lads – the prospects, at the age of seventeen, of actually leaving, and going to somewhere like London or Birmingham where you know nobody – where there are no relations, where you don't speak the language, you don't know the customs – it's all alien, completely alien to you.

It's even more difficult to leave when the jobs may not be there. You just don't know how you do it. What are the mechanics of actually getting there, and where do you live, and how? Mates, and the community, are a thing you're born to, and actually defying that, *even* to get out of the pit – it's just a prospect which is not on.

Of course, now in the days of massive unemployment and abandoned industry all around us, the pit is a real source of employment again. Lads are desperate for work. Nearly every colliery in Doncaster records waiting lists 300 and 400 strong before the Board is forced to close the lists. Some of the lads have been waiting more than two years, which shows the sheer desperation of the situation.

At the same time, the purge is on. In years gone by, the worker pleased himself when he went to the pit; he only worked for money. When he had enough, and no special needs were outstanding, he stayed at home and enjoyed the sun. There is the now famous story of the gaffer threatening

Bentley Colliery with closure because of absenteeism, demanding of the crowd, 'Why do you only work four shifts?' Back came the response, 'Coz we can't live on three!'

Another mate of mine at Hatfield, a sunny afternoon when he was due to be on afternoons, went in to see the undermanager. 'Gaffer,' he says, ''a've got ti be off work for a while. Wor lass is gaanin ti have a baby!' 'Oh, very nice,' says the gaffer. 'That'll be OK I suppose. By the way, when is it due?' The lad looks at his watch, 'Well, after a'h get home, about nine months time!' In these times, taking advantage of the unemployment, the Board is challenging the ancient custom. Sackings for absenteeism are sweeping the area, and men once again go in fear of losing too much time off work. Even 'the sick note' does nothing to cover you. A record of sickness allows the Board to claim you are not fit to be employed in the industry – and out you go. Unemployment, it seems, is the only escape nowadays.

⑨ RANK-AND-FILE RESPONSES

The annual Miners' Gala in Durham is a remarkable monument to the very special class spirit of the British miners, with its various traditions of militancy and Methodism. On the second Saturday in July, each colliery marches behind a silk banner that may bear the image of anyone from Lenin to Harold Wilson – a scene from the 1926 general strike following one from the Bible. Speakers for the event are chosen in a local union election, and the invitation to speak is a great honour. (No one is quite sure whether Prince Kropotkin, who once addressed the gala, was invited because he was a prince or because he was an anarchist. It's likely that he received votes from more than one source.)

The gala is also a good forum for a major policy statement. Traditionally, a Labour prime minister or a high Labour cabinet member has a prominent place on the programme. And invariably the ranking party member is introduced as 'my friend Tony' or 'my friend Michael' – the trappings of power left elsewhere. Facing 10,000 miners and their families, it is clear to the ministers – and to the BBC cameras – where the balance of power resides. The Labour leader simply tries to tap this power – or restrain it.

In 1977, however, the featured speaker was not prime minister James Callaghan. It was the NUM's Yorkshire area leader Arthur Scargill, Callaghan's most outspoken critic among the union leadership. (It was Scargill's first invitation from the moderate Durham miners.) In June the NUM had sent the government's pay policy skidding when it approved a wage demand of some 91 per cent. Nor was Callaghan likely to forget that it was the miners who brought down the Heath government in 1974, defying the Tories' last attempt to impose wage controls by calling their second national strike in two years.

Of course the miners would remove a Labour government with much less pleasure. But the gala invitations made it clear that rank-and-file miners did not feel Callaghan had held up his end of the deal as a Labour prime minister. They were tired of the Social Contract which obliged the trade union movement to accept an incomes policy which allowed inflation to exceed wage increases, and held Callaghan responsible for what felt to them like a sharp decline in living standards.

Although not invited to speak, Callaghan was on the podium. He looked increasingly owlish as Scargill, turning to him several times in the course of his speech, heaped on abuse for the government's betrayal of the trade union movement. 'No wonder the CBI (Confederation of British Industry) have welcomed with open arms Mr Healey's statement on limiting wage increases to 10 per cent at a time when profits are running at a record 47 per cent in increase over last year.' Scargill assured both Callaghan and the Durham miners that the NUM pay claim was more than a public relations ploy or a sop to militants. In closing, the Yorkshire leader spoke beyond the local miners to a wider audience. 'I want to urge the whole trade union movement, and in particular my own union, to ignore the advice and pleas of the government for further wage restraint policy.'

After more harsh speeches, followed by some that took a conciliatory tone, Callaghan asked to address the crowd. The master of ceremonies introduced him in the hard regional accent that renders the name 'Kaligan', skipping the part about 'my good friend'. But despite a jeering audience, the prime minister hit his message hard. 'For the sake of the country – and miners are part of the whole national family – the government's first priority is to get down the rate of inflation which was destroying us three years ago. That is the objective we have set ourselves, and the tide is beginning to turn.' He accused the miners of demanding 'the lion's share' and trampling on the needs of 'the weak and the sick and those whose bargaining power is not so strong'.

Unlike his Labour predecessor Harold Wilson, Callaghan rarely took on hecklers. But this day he engaged in sharp repartee. 'He who laughs last laughs best,' he told the crowd, to which one miner responded, 'You can afford to laugh.' 'Jeers are what we hear from the Tory benches every day,' Callaghan replied, 'from people who argue that the government cannot succeed in bringing down inflation.' He concluded on a more ministerial note, looking straight into the television cameras: 'Next year the miners' claim will be £200 a week, and the next year £300 a week. Is the miner worth it? Certainly he is, provided he is paid in real money, that is what we have got to do. Therefore, let no one think that I can be carried away on this particular issue. What I am looking for is the real standard of improvement and advancement for our people, not easy slogans.'

Callaghan had won the crowd back. Not entirely, of course, but he was Labour prime minister and he had waited his turn and boldly faced the opposition. He had struck the right chord by appealing to miners' sympathy for and solidarity with the less powerful. You have your sectoral interests, Callaghan was saying, but I have the interests of all working people at heart.

The miners did not bring down Callaghan's government but they remained a powerful force to be reckoned with. Even the indomitable Margaret Thatcher backed off from a confrontation with the miners over pit closures in the first year of her government – and the firmness of the miners' resistance to any encroachment serves as a reminder to any unwary – or hell-bent – government that Britain is still a class-torn society.

Nobody wanted coal. The miners were a legacy of the past and a great many people down south had not the slightest concept of the extent of the coal industry in Britain. To them it was as rare and colloquial as the shire horse ploughing a field. With this as a background, the union leadership preached endlessly that any action to win wage rises or concessions in hours or welfare would simply hasten the closure programme.

However, things had been changing so gradually, that many people had not noticed that the ground they occupied today was not the same as that they occupied yesterday. It would be decisive if ever the push came.

Many factors went into this gradual change. Perhaps the most significant was the signing of the National Power Loading Agreement (NPLA). For the first time, miners all over the country would have the same rises according to their grades, a coalface worker in Nottingham would be earning the same rate as a coalface worker in Scotland or Wales. Piece rates were abolished, and it would be with one voice and one effort that the union would approach the Board as a single employer at national level, rather than a series of local bosses reflecting local strengths and weaknesses of the various union regions. Different rates would continue until the lowest had year by year caught up with the highest, but henceforth we would have one job, one rate, and one union to fight for it.

In the 1840s when the Geordie agitator pitmen toured Durham and Northumberland spreading the union and industrial action cheek by jowl, they used to shout to the crowd, 'D'ye a'll agree to form the union?' Back would come the shout, 'Whey—aye!' The next cry was, 'Whey what's the forst thing ye dee nuw ye've formed a union?' Back would come the yell, 'Strike!!' So it was with our unofficial movement, which lay behind the 1972 strike.

A great unofficial strike movement swept the coalfields in 1969. At first we were undecided as to which part of our oppression should be challenged first, but as the movement spread, the protests united into one solid chorus of discontent. While wage negotiations were still in progress, the Welsh, Scottish, and Yorkshire miners downed tools and struck for a reduction in working hours for the surfacemen who still worked hours imposed after the defeat of the 1926 national strike. This strike of 1969, which involved 150,000 miners from 130 collieries, united numerous rank-and-file papers, journals, and unofficial organizations around a common tactic and programme of demands. The strike failed to immediately win a reduction in surfacemen's hours, but it did ensure that we received the biggest wage award in the miners' history.

The 1969 strike was the cornerstone of all subsequent strikes. It swept aside the old union leadership and gave birth to a new, 'left' leadership. It caught the Coal Board completely unawares and, within a few days, started to bite deep into their revenues. Most of all, it woke the mass of miners up to something the militants had been preaching for years, namely that the country was still very much dependent on coal and that the miners were still in a strong bargaining position. Once the argument was proved, the perspective of the rank and file began to change tremendously.

The unofficial movement, led and orientated in a hundred different collieries by many different forces, was able to maintain its push in a consistent direction decided by internal democracy. Perhaps the two most important points on any one of the militants' programmes were: (1) a simple majority was needed to call strike action, and (2) all officials were subject to instant recall and annual election. We are still fighting for the second point, but number one was won hands down. Previously, we had to secure a two-thirds vote in favour of industrial action to call an official strike, but now we need only 55 per cent. This is in itself a major breakthrough. Once the rank and file had the means and the notion to progress, they set off like the clappers to capture lost ground.

Of course, peoples' memories of the 1969 movement are coloured by where they were standing at the time. The scene, the background to that movement will differ according to your involvement; the level of that involvement, whether as an area leader, a minor branch official or a raw rank and filer; and, of course, the locale from which you operated at the time.

There is a tendency among some to see everything in terms of personalized conspiracies, or a series of personality coups, the engineering for which takes place among a small group of conspirators 'in the know' in smoky backroom pubs of undisclosed location, but this just wasn't the case. Although I, like many another militant, was out on the streets, round the villages, addressing meetings, going on marches and spending long nights in endless duplicating of agitational leaflets, few if any of the people said to be at 'the head of the coup' were known to me and doubtless vice versa. It made me wonder if there were not at least two conspiracies taking place at the same time, one in which the rank and file engaged and the other which took place in influential election coups and platform victories at Barnsley and elsewhere.

For most people, the miners were reborn in 1972. It was a renaissance of general class militancy muffled and shuffling for 40 years prior. The year 1972 is seen as the turning point in miners' history – a movement which posed for the first time *national* industrial action which had been long gone but not forgotten. It might be added that notwithstanding 1926, the union still carried the marks of its federated past and was riddled with parochial and divergent traditions. Militancy in the coalfields, though often heroic, had almost always been un-co-ordinated. It was in building a platform, through inscribing on the banner 'national industrial action', through learning again the language of militancy and starting the march of *action* that 1969 framed its unique contribution to our recent history.

It was at once a multivariety phenomenon and it took the form overall of a loose amalgam of 'leftist' and 'militant' leaders – sometimes with their own documents and leaflets – along with young revolutionaries very influenced by the general high tide of the student-worker revolts sweeping Europe, and the native sub-culture of tin pan alley revolutionism portrayed in pop music and festivals. There was a profusion of leaflets, meetings and journals. In Doncaster, for example, we distributed the *Mineworker* with a base not only in Doncaster but a good-size readership in Wales and Derbyshire. Then there was *Link Up*, which we might call left social-democratic and militant, comprising leftist Labour Party members and the leaders of local parties. Also, there was the *New Link*, voicing many of the local branch leaderships. We must, I think, add to this the products of the trade union day release courses from Sheffield, Leeds and Derbyshire. There were discerning, educated trade unionists able to orientate the growing discontentment and destined in many cases to fill local lodge positions. They were, I believe, an important element.

Undoubtedly, one of the prime instruments which gave structure and form to the unofficial movement and took over to all intents and purposes the running of the coming strike, was the miner's panels. These panels functioned far beyond their stated remits, and to many appeared as joint shop stewards' committees on the one hand, or quasi-soviets on another. These bodies are local assemblies of all the branches in each designated NCB area, originally established for consultation on matters of joint concern and rather toothless, but they rapidly passed over into organs of policy and action, responding rapidly as they did so to grass-roots frustration.

A major challenge to union leadership was launched in

practically every district of Derbyshire, Wales and Yorkshire, and loosely co-ordinated. Four panels were subsequently to lead to a coup in lodge positions all over the country and a revived interest among large numbers of the general body of the NUM.

A new feeling was sweeping the coalfields. Mass meetings at area and village levels saw new faces voicing an old message appearing on the platforms. It was little Derbyshire who raised the first flag and organized a demonstration in London demanding justice for the surface workers. It issued a clarion – though quite unorthodox – call to all areas to rise. It was a great day when, confused and disorganized though we were, we assembled in London: Welshmen, Yorkies, Scots and Derbyshire men. The wide-flung coalfields were together, and of the same mind in their anger and disgust, we felt a national union – and confident. It was a rallying thing in itself, to stand in the throng and hear the diversity of accents, the standing stones of labour history. With the power of agitation surging in our veins, we headed back to the districts, determined to turn the system on its lug.

Whilst a heated council meeting was taking place at Barnsley offices, thousands and thousands of militants swarmed around outside demanding action! Action! And as the TV cameras swept through the crowds, the powerful lyrics of 'The Miners' Lifeguard' were heard from the close-packed ranks:

UNION MINERS STAND TOGETHER
DO NOT HEED NO BOSSES TALE
KEEP YOUR HANDS UPON YOUR WAGES
AND YOUR EYES UPON THE SCALE.

Nationwide, while wage negotiations were still in progress, the Welsh, Scots, and Yorkshire miners downed tools and struck in demand for a reduction in working hours for the surfacemen. In the Durham coalfield, small pockets of militants tried to generate a movement and actually left collieries idle here and there; pickets composed of transferred Geordies had made the trek back home in an attempt to pull out their mates. The Durham Miners' Association (DMA), however, true to form, was 'outraged' and held disciplinary meetings and threatened to withdraw the credentials from lodge officials.

The 1969 strike brought 130 collieries to a halt and 150,000 miners to their feet. Within days, the strike started to bite; few firms had stockpiles of coal. Soon factories were laying off men and the truth was emerging. Nuclear energy bollocks! Coal was a powerful commodity. We had a strong bargaining position; the pitmens' labour had real value, and from this day on we would be paid fairly for it or let them burn turf. The strike had a huge impact on the negotiations and wrung from the Coal Board the biggest rise in our history up 'til that time.

The strike was symptomatic of other things, too. Simultaneous to the mass strike movement, there was a parallel constitutional battle. This involved the age-old battle to change the odious rule relating to national official strike action. The rule necessitated a two-thirds majority vote by the membership cast in the affirmative for a strike to take place. The requirement was as impossible as it was absurd, given the number of retired members, long-term sickness men, and 20,000 affiliated clerks. The rule, in effect, necessitated a 100 per cent affirmative vote by working miners.

As a result of the 1969 movement, the old strike rule fell and a new 55 per cent affirmative vote was instituted. Without it, the successful vote of 1972 and 1974 would have been to no avail. This constitutional victory was a major breakthrough in our ceaseless battle with the NUM bureaucracy.

In retrospect, it must be said that the strike (initially at least) did not win a reduction in surfacemen's hours. That was to come later. What we did was to put it fully on the agenda, and in the process, won the biggest wage rise ever. More importantly, we had flexed long-tired muscles, had demonstrated that militancy worked and that we had a powerful weapon at our disposal. We regained our self-respect. In the locals, we swept away much of the old leadership; we precipitated the rise of a new leftist current. We laid the mark lines for what in future years would be called the storm troops of the Trades Union Congress (TUC).

The 1972 national strike was a high point in the miners' history. The rank and file established its control of the strike early on. But many of the more conservative areas remained true to form. Places like Durham and Northumberland voted heavily against strike action. In fact, we just achieved the simple majority in favour of a strike in a national vote. After all, this was the first national strike since 1926. However, once the strike movement got under way, the workers really came to grips with the state and pitched in to the battle for all they were worth.

The Tory government was fully committed to breaking this strike. Just prior to the miners' action, many sectors of the working class had been knocked back by the collective efforts of the government and the capitalist class. The government had publicly urged all employers (as if they needed any urging) to hold out against wage demands; it also declared that employers who paid wage increases were equally responsible (along with the trade unions), for the economic ruin of the country. The government itself was absolutely determined to hold out against the workers in the state sector, especially the miners. Indeed, the miners were breaking the new anti-strike Industrial Relations Act, and to many it was touch and go whether or not the Tories would use the law's penal sections against the pitmen.

But the workers had many strengths in this big 1972 strike. From the word go, the strike was properly co-ordinated at regular union branch meetings in which the rank and file could make their views felt. The local leaders, particularly in the highly influential Yorkshire and Scottish coalfields, had their eyes set on national fame and coming elections. These men went out of their way to parade their militancy and to impel the men to greater and greater feats of class courage. A new-found power surged through the union, solidifying union officials with the rank and file. For the first time in living memory, the union really was 100 per cent solid behind the struggle.

The pickets were absolutely unprecedented in their mammoth size and determination; they were also unusual in their intransigence and in the success they achieved despite real police brutality and the use of scab truck drivers who murdered at least one comrade and injured countless others. The pickets were heavily supported by the miners' wives, whose determination added a new facet to the action. Scab drivers, some of them armed with starting handles and iron bars ready to meet a line of pitmen were in many cases nonplussed by a just-as-sure line of miners' wives and daughters. The miners had moved into battle with their whole community actively engaged. Countless workers from other industries, particularly truck drivers and railwaymen, stood shoulder to shoulder with the miners. The flying pickets, which evolved as a more sophisticated version of an earlier practice of spreading strikes from pit to pit, took on huge dimensions; they travelled from one part of the country to another, pouncing on the hundreds of unsuspecting coal and oil depots. They went so far as to form a navy in the Thames to prevent barges from landing at the coal wharves.

Actually, the story of the flying pickets is quite funny. Arthur Scargill always gets the credit for 'inventing' the flying pickets. Our old delegate, Tom Mullany, very proud of his involvement in historic actions, stood up in the council chamber and said, 'Well, I don't know if Arthur can remember as far back as I can, but I remember flying pickets coming to Hatfield in 1919, they came in a steam bus from Askern and they run over Amos's dog!'

The nature of the miners' pickets was such as to allow the greatest possible access to revolutionary ideology. For one thing, vast numbers of miners were housed in universities near the sites they were picketing; they lived and slept on the campuses and, of course, talked there. Moderate students were beginning to see the reality of the class war illustrated in the tales of the miners. The reverse was also true; revolutionary students had an unprecedented opportunity to talk and discuss with miners at meetings, on pickets, and most importantly over meals and over pints of beer in the student union bar. The same was true on the pickets themselves, where large numbers of students joined to stand out in the cold and rain at all hours of the day and night shoulder to shoulder with the miners. What a contrast to 1926, when the university students had driven scab trains and buses to ruin the general strike. Now they were halting scab trucks and were joining demonstrations in support of the working class: what a symbol of middle-class disenchantment with the capitalist system.

Metalworkers, labourers, factory workers of all kinds joined with the political forces of the left to hold back the trucks. Countless socialist and revolutionary papers, leaflets and journals bombarded the pickets and, in such circumstances as produced long, long periods of sitting and talking, these left publications never failed to stimulate discussion, dissent and, most often, hardened agreement on the basic points.

The flying picket reached its absolute crescendo at Saltley Gate coal depot in Birmingham. After weeks of escalating pickets, with both miners and police pouring more and more people into the struggle, the situation was beginning to resemble a medieval battlefield black with bodies of men charging and countercharging. Then came the climax. When the news spread through the Birmingham factories that some

5,000 police were in action at Saltley Gate, 20,000 metal-workers downed tools, factories closed, and a great army of Birmingham workers marched to support their miner comrades. The sight of all these workers, flags and banners flying, marching determinedly to the old refrain 'Solidarity Forever' sent the pitmen wild with excitement. The chief of police took one look and rapidly made up his mind: the coal depot would close – it was a hazard to safety and public order. The closing of the vast plant was a major victory for the miners and for the whole of the working class. If the Tories wanted confrontation, the miners were the ones to give it to them – and to come out with the laurel leaves!

Once that big victory had struck home, the Tories resolved to drop the confrontation with the miners but to continue their attack on the rest of the class. They decided they would make the miners a 'special case'. And because the miners were a 'special case', they had a 'special minister' to look into things: so they formed the Wilberforce Enquiry. From the first moment of this invention, there was never any doubt that it would come up with findings in our favour. Wilberforce was a rabbit out of the hat, the same hat which had produced the previously unheard of 'solicitor general' to get the Tories off their own hook when it looked like the dockers would catalyse a general strike to get their workmates out of Pentonville Prison earlier in 1972. The Tories were forced to settle almost the full claim of the miners in order to take the steam out of the class struggle.

Although the wage claim wasn't met 100 per cent, the miners won a fantastic victory. For the first time in years, the miner felt like he was getting back his position in the wages table. But more importantly, he was regaining his old pride as a class fighter in the fore of the working class. Political meetings and demonstrations of all kinds on all issues suddenly saw very large contingents of miners, whereas previously only the few union branch stalwarts would appear. In the revolutionary groups, quite mature miners started to be seen speaking on left platforms and writing in journals. The wage victory had also meant an advance in class consciousness.

It is worth mentioning the simultaneous advance of the left within the Labour Party. The Labour Party rightly or wrongly is seen as part of the trade union movement and vice versa. Movements and trends in either realm are usually reflected quite heavily within the other. The same was true on this occasion. Coupled with the tremendous victory of the miners and certain other sectors, the left inside the Labour Party started to gain ground. A whole programme of nationalizations – of land, banks, oil, etc. – was passed at the National Conference. More important was the really big vote for nationalizations of one hundred monopolies without compensation and under workers' control. While the motion was not actually passed, the vote in favour was very significant. The character of the National Executive of the Labour Party also has changed quite a lot; while it could not be called 'socialist', it was certainly the most left National Executive Committee ever elected. With one or two notable exceptions, the left began purging the right, expelling particularly odious right-wing MPs and local councillors.

This is a process well under way in current political struggles. The 'weeding out' has led to the formation of a right-wing 'labour' grouping in the shape of the Social Democratic Party; other entrenched right-wing sections of the party seem poised to leave if they lose the battle for the heart of the membership. A last-ditch battle is going on with some casualties on the so-called 'far left' being refused membership or endorsement of candidature for parliament or council seats. What seems irrefutable is that the process of leftist struggle within the party and the dedication to oust the right social democrats from positions of power received a tremendous boost following the 1972 victory.[3]

In 1974, despite the partly successful efforts of union officials to win control back from the rank and file, the strike which brought down the Heath government revealed, once again, a really brilliant flash of class understanding by the miners. When the battle was all but won, just before the elections, a concerted effort was made by the parliamentary Labour Party to get the miners to 'suspend' the strike while the elections were taking place, just to show that the workers really do believe in Parliament and the niceties of bourgeois democracy. Although sections of the NUM leadership were prepared to concede on this point, the answer from the rank and file was loud and clear and too obvious to be ignored: 'No bloody chance!' Elections or not, the miners wanted justice!

A Labour government was returned by the elections, winning on a platform of nationalizations, repeal of various authoritarian Tory laws, a settlement with the miners, and a return to work for everyone on a five-days-a-week basis. The new Labour government allowed the National Coal Board to negotiate freely with the NUM. Their eventual offer was based very much upon the findings of the 'Relativities Board', which were published just after the election.

Though it contained great elements of victory, the offer failed the lower-paid workers. Faceworkers were offered £45 per week, as had been demanded. But underground workers were offered £36 per week instead of the £40 that had been demanded. And surface workers were offered £32 per week instead of the £35 that had been demanded. The National Executive Committee (NEC) accepted because it was more concerned for the Labour Party than for its own members. To gain passage of the settlement, the NEC had to violate its own rules. The NUM rule book requires that a ballot be taken of the membership to decide whether the members accept or reject a contract offer. That would have given the rank and file time to discuss the issues during the weeks it takes to prepare a ballot. It would have allowed the revolutionary and left groups a chance to call meetings and influence the workers. So, the NEC committed its final act of sabotage. It demanded a vote of the branches, which required only days to prepare. Only the national, area and local bureaucrats had occasion to brief the members on their arguments for acceptance. The agreement was a low price to pay for a great movement which could very well and very easily have wrung every penny out of a Labour government that was almost duty-bound to settle with the miners.

But, despite the NEC's sabotage, the miners won a great political victory. They threw out a Tory government because their class courage was too hard a nut to crack. They won a great many, if not enough, of their economic demands in the final settlement. The elements of sell-out are the sole property of the bureaucracy. The elements of victory belong exclusively to the determination of the miners and the British working class.

More recently, the militancy of the response has fallen off. The only exception to this was the recent reaction to a new scheme of mass closures. Fifty pits had been scheduled to close. At first it was denied, but then reluctantly admitted. With the dole queue as a whip master, the areas rallied, with no exceptions, and a great unofficial strike movement jumped the gun, as the areas planned an all-out national strike. Within hours of the time set for an all-out stoppage, the Board, manipulated by the government, withdrew the plan. The old unity of the miners was still there if they had common cause.

Only time will tell whether the unity can be brought to bear in our struggles against the incentive productivity scheme. Without a doubt, the lack of militant response to recent wage claims has been the effect of the bonus system in making rich men of some and poor men of others. When some men can earn bonuses twice as high as their wages, and others have only their flat wages, it is clear that there will be no harmony of interests.

Certainly it is a continuing injustice. The Christmas of

1981 brought the real wages back to the men and their families. Because of the holidays, the Board decided to pay out only the flat rate, no overtime, no bonuses. The real rate of pay was there for everyone to see and make a lie of the press statements of what miners earn. We see here the grand sum of £75 take-home pay for faceworkers. True, with bonuses you *could* make more, but even with all the will, sweat and even blood in the world, some faces will not produce bonuses for these men working the very hardest underground.

The areas' resolve to resist the scheme weakened one by one as the Board threw their money at lodges to get the system accepted. The old pragmatism of the moderate miners broke through. 'If there's money to be got, lad, we're having it; that's what we work down pit for.' This was to be no 'productivity' scheme the Board assured us; it was not coal *production* but *effort* which would be rewarded. Consequently, men in difficult seams would have as equal a chance of getting money as those in good geology.

Now that the courtship period is over, it is clear that the nineteenth-century yardstick, 'no coal – no wages', applies again. The pits with level, high and clean geology, equipped with all that modern technology, can come up with coal pouring out like a great river. These colliers are well satisfied, feeling they are kings of the wages league at last.

Meanwhile, in April of 1982 at Hatfield Colliery, the other side of the scheme showed its true face. H26s unit is holed, with a cavity big enough to take three double-decked buses. The men standing on top of the chocks stand in the very jaws of hell. They try to timber the hole in much the same fashion as their very earliest mining forebears, with a wooden trellis of half tree trunks and wooden blocks, layer upon layer, until the structure reaches the trembling roof and is wedged tight with wooden pinners. It is an attempt to control the earth, at least temporarily. But before she submits, great slabs will suddenly flop out and demolish the structure, smaller lumps loose themselves silently from the dark roof of the chasm and strike the miners below on the head, shoulders or arms. Great weight bumps 'flush' the goaf or waste areas behind the chock line, and clouds of dust add to the fear eating at their bellies as their eyes dart like those of an animal with every life-saving wit aroused and on call.

In the same month, Hatfield's HR2s unit carried two vast holes on the face and stretching back over the chocks. If you can imagine looking over the end of a cliff, thirty feet deep, you will get a picture of what it is like to look up into one of these holes. Those great boulders above your head are being held up by nothing at all other than the whim of the earth itself.

In this case, the holes were too vast and men had to try and 'dowl' the face with a kind of plugging system, in the hope that the next few cuts of coal would not bring the roof in further. Two men using a boring machine must stand beneath that vast chasm and bore holes in the coal. Another miner stands with them, head back, light shining up into the cavity, waiting for the least sign of movement in the top so he can give a split-second warning to his two mates to drop the machine and dive for cover. He stands, one hand holding his helmet on as he leans his head right back, the other on the shoulder of his comrade's exposed back so that the brief warning is that much more speedily delivered. This is the nearest thing to 'eyes in the back of your head' that I have ever seen.

For all this toil, danger and effort, these coalface workers received the princely bonus of £1.90 per shift. No coal – no pay.

What turns the knife even more – and there's no jealousy in this, simply a sense of injustice – is that other people who have never even seen a mine get bonuses ten times greater. The incentive scheme gives to some classes of workers, 65 per cent, 50 per cent and 40 per cent, respectively, of the 100 per cent coalface workers' bonus. This means that at pits where coal is easily won, the canteen staff, the clerks, and the juveniles, for example, receive a higher take-home pay than those men in the belly of hell at Hatfield's holed faces. The white-collar workers down in London, although only on 40 per cent of the area's average bonuses, manage to take home more bonus payments in a day than the coalface workers at Hatfield take home in a week.

This is the injustice the miners of the present must address themselves to in the coming years.

EPILOGUE

Now, in the month of July 1982, it looks as though the whole gamut of closures and mass run-down are looming again. After the last closure list was withdrawn under the weight of a certain national strike, an uneasy calm descended on the coalfields. But there were no wild celebrations of victory when the threat was withdrawn; we knew in our heart of hearts that this was simply a breathing space.

In the meantime, although the list was withdrawn, manpower has been running down like the industry's lifeblood trickling away, and redundancies offered to old men who, with 40 years in the pit and a chest full of dust, jump at the chance for a lump payment and a chance to get out early. Many of the men going out under redundancy schemes are in fact long-term sick and will benefit greatly by being included as 'redundant'. The purge on absenteeism, a block on virtually all juvenile and adult recruitment – all of this has slashed and slashed again at the strength of the industry. Something like the equivalent of ten or fifteen pits have been lost in terms of reduced manning.

Then the threat came again to make the trickle a flood. Snowdown colliery in Kent was threatened with partial closure over the next three or four (or more) years. A new policy began to emerge. Investment in coal considered 'marginal' would be withheld and only rich seams would be developed. In the meantime, that marginal capacity would be shed and jobs were to be taken off us. It became apparent that a great many pits could be run down or put into suspended animation in this way with a consequental job loss easily equivalent to the original 'hit list'. Jack Colins, leader of the Kent men, felt as though this was a fight for the whole of his area, not just Snowdown, and old fears began to seep up from coalfield to coalfield that the Thatcher government, flushed with its jingoistic blood-letting in the Falklands, had chosen its time to fight round two of what looks like another elimination contest.

On Thursday 24 June, the National Executive Committee of the NUM were set to meet the NCB under the formal heading of Investment Prospects of the Board – Future of the Industry, at the Board's grand offices at Hobart House in London. Urgent answers to highly specific questions were needed and needed fast. A mass lobby formed up outside from early in the morning, and suddenly it was 1969 all over again. We had picketed outside this building then, the same concerned faces, the anger just kept in check, the accents of Geordie, Welsh, Scottish and Yorkshire, the pinstriped occupants arriving in taxis and pushing through the crowd, the female office staff fearful and perhaps contemptuous of those foreigners who had upset their day. The National Officials arrived one by one and were urged 'no closures, no run-downs, no sell-outs, defend our jobs'.

As officials of the Board entered, we were at a loss to establish who was the tea boy, who was the desk clerk and who were the big wigs, so shouting was muffled and uncertain. A row of uniformed commissionaires stood guard on the revolving doors, and the other doors were bolted against our entrance. We had a feeling that enough wasn't being done, that this was a half-hearted affair. Then it started raining and the cops wanted to move us down the steps and on to the pavement. Groups were being herded about. Suddenly Colins shouted, 'We're going in here lads, come on!' and a rush was made for the revolving doors. Half a dozen miners wedged in each section of the door and slowly edged it round against the combined efforts of the commissionaires. Progress was slow since the Kent men had wedged their banner poles into the doors as well. Then a commissionaire or two fell, the door moved faster, another commissionaire got his foot stuck in the revolving door, and he started to be pulled round to the outside, shouting, 'Me foot's fast, me foot's fast.' So Colins shouted, 'Well, take the fucking thing out then!' A determined but jovial sortie broke out and soon the lower reception area was filled with miners. The Kent men erected their banner, the clerks ran into their offices and bolted the doors, shouting, 'They're in, they've got in' as though the Martians had landed.

Then, led by Jack, off we all marched down the corridors into the room where the NEC was meeting. Arthur Scargill looked surprised as the men flooded in, then slightly shocked as they kept flooding in and flooding in, maybe 400 or 500 strong. Then the Kent banner came in and he looked rather helpless—so I shouted, 'It's OK, Arthur, the band's coming next!' We had made the point to our own leaders if no one else, that we were as determined as ever to stop run-downs and closures. Colins told the NEC that our presence was no disrespect to them, but we felt our determination had to be shown. After a bit more crack, we withdrew to the reception area.

After the NEC meeting, they waited on the Board to come in. Then Arthur came out to us and quietly told us that the Board wouldn't meet us while we were in the building and would we go back outside again, amidst shouts of, 'They didn't do it like that in the Lenin Shipyard.' But Arthur explained that the Board, in his view, were not going to lay their cards on the table and had nothing to hide their treacherous blushes with and were using our presence as an excuse for not meeting – also that it would be used as a propaganda item by the Board if we didn't move. He assured us that he did not think we had long to wait outside.

After we got outside, about fifteen minutes passed when

out came Scargill again, telling us that the Board were not prepared to tell us anything about manpower and closures and would only talk in general, in no helpful terms. The NEC would be meeting to recommend industrial action to the Conference, unless the Board told us their plans and withdrew any designs they had for closures and mass forced redundancies and transfers. And that was how it stood.

This book closes with no certainty for the future, as, in fact, it began. So for the individual miner, so for the unions – a knife's edge of danger for today and uncertainty for tomorrow, but the conviction that every threat and new repression will be met blow for blow.

GLOSSARY

airstream helmet

This is a modern invention, designed to cope with excessive dust. The helmet has a small fan which pulls air up the back of the helmet through a filter. Then it redirects the air down the front, while a screen keeps the air away from the face. The trouble is that the helmet is heavy, and also it tends to fall off if you have to crawl around.

back-bye

The areas away from the face.

backripping

The facerippers advance the tunnel behind the coalface. However, after a while the weight of the earth comes down on to that tunnel and starts to close it up again. What is required then is that a team of backrippers come to that area and start ripping it again. They take out the old crushed arch, extract enough of the roof or floor to make the tunnel the right height again, and then put in a new arch.

beltmen

These are the men who install the conveyor belts which lead out the coal from the mine. They also patrol in teams various areas of the mine, repairing and changing worn belting, etc.

boom ripper
This is a modern cutting machine used in the making of roadways.

bow, crown, centre-piece
The gate or tunnel is advanced first by extracting an area of stone, then this area has to be supported. This is done by constructing a steel skeleton in the shape of an elongated semicircle; this we call the 'ring' or 'arch'. The arch is made in three pieces – the bow or crown and two legs. The bow is the first piece to go up. It is temporarily secured on two iron bars protruding from the last completed arch – these we call the horseheads. Once the bow is up, we cover it with corrugated iron sheets, and pin these to the bow by wedging them between the sheet itself and the roof overhead by means of wooden wedges. Once this is done we can bolt the legs to the bow, thus making the structure free-standing.

cage (chair)
This is the metal structure, open, and enclosed with a kind of mesh grillwork, in which the miners 'ride' up and down the shaft to and from the pit bottom.

cap lamp
The battery lamp, strapped on to the belt and worn on the helmet.

caunchmen
See ripper.

(hydraulic) chocks
This is the modern system of roof support. It is like a small vehicle with telescopic legs under a metal canopy. These legs can be lowered, the chocks moved forward, then the legs pumped up to hold the roof again.

chocking, ramming, and **grading**
These processes involve advancing the chocks, ramming the face chain forward, and checking the cutting horizon so as to keep a clean cut of the coal, neither cutting into the stone floor, nor leaving coal bottoms behind.

drivage, scour, drift heading
Various names given to the tunnels which are 'won' through the solid strata, rock or stone of one sort or another, in order to get to a new coal seam, or else to connect up one roadway to another.

exchange
This is the control centre on the surface. It is in touch with all the faces in the mine and co-ordinates reports on production, hold-ups, etc.

face chain
This is the conveyor which runs along the coalface and carries the coal which has been cut off the face on to the stage loader chain, then in turn on to the conveyor belt out of the mine.

fire damp
Methane, highly explosive gas which comes off the coal.

flame lamp (oiler)
This is the miner's 'Davy' lamp, invented to burn without igniting the methane in the air. It is used exclusively now as a means of testing for gas, although it was used in the past as a 'safe' means of illumination.

gaffers
NCB managers and their agents.

gob (goaf)
The worked-out area which is allowed to collapse, behind the chocks.

handfilled
Faces or lips (rips) from which the coal or stone is 'filled' on to the conveyor belt by use of the shovel without the aid of machinery.

hewers, fillers, colliers
The men who work on the coalface actually getting the coal. The hewers and fillers traditionally shovelled the coal and felled it with a pick and shot-blasting.

hoggers (banickers)
Short pants used by working miners, like old football shorts.

keps
Four iron bars which drop down from the side of the shaft, on which the cage comes to rest.

knackered
A Geordie expression meaning totally exhausted or worked out.

lids
Wooden block about six inches square used to tighten wooden and hydraulic props.

lowse
This refers to 'knock-off' time. The term probably came from the days of the old hempen ropes. The chaldron would be 'lowsed' or loosed off in order to allow the men and boys to ride up the shaft on the rope in the 'arse

loops', at the end of the shift. (The chaldron, or corf, was a kind of wooden or woven 'tub' used in the eighteenth and nineteenth centuries.)

main gate
Main tunnel. *See* ripper.

marra
Your work mate is your marra. It was a Geordie expression at first, but the term has drifted south with the Geordies and is now common in South Yorkshire pit talk.

methanometer
A rubber meter for testing for the presence of methane. It operates off a battery.

neuks
The corners or ends of the face.

onsetters
These are the men in charge of the cage at the bottom of the shaft. They signal for the men or material to go up the pit.

pack hole
These are the solid pillars of stone built at the face ends to hold up the tunnel.

paddy
This is the name given to the various sorts of manriding equipment underground. The origin is obscure, but is said to relate to the numbers of Irishmen employed in some Yorkshire pits in the old days. Two principal types exist, firstly the set of coaches which are drawn along by a locomotive in much the style (if not the comfort) of a

surface train. The other is a set of coaches which are
drawn along on a rope which winds on and off a big drum
according to the direction it is going.

pinners

This is a wooden wedge about one foot long, used to
tighten up wood chocks and wooden trellis work used in
timbering cavities and holes above the tunnels.

poke (bait poke)

This is an old army bag, usually, used to carry the bait
(any food eaten underground). Poke also refers to any
small bag.

pullers, drawers

The men who withdraw supports behind the face to let
the worked-out area collapse, and who advance the face
chain or belt forward into the new cutting track. The coalface goes forward all the time as the coal is withdrawn, so
the machinery behind it has to be advanced to keep the
process going.

putters

These were the lads who brought the tubs in to the fillers
to be filled and brought the full ones out again. They
either did this by hand, or rather by head, pushing the tub
in and out, or else with the aid of a galloway or pony.

ripper, caunchman, stoneman

This is the coalface worker who works at either end of the
coalface advancing the 'gate' or tunnel. If one can imagine a
soccer goalpost lying down, the top bar is the coalface and the
two 'legs' are the tunnels or gates in to that coalface. While
the colliers or fillers advance the coalface itself, the rippers
advance the tunnels or gates. Up the tail gate go all the
materials for the face and out of the main gate (mothergate,
or mullergate) comes the coal on the conveyor belts.

sleepers
These are the same as the sleepers on surface railways, the wooden batons which are placed at regular intervals under the lines.

stable machine
The stable is the tail-gate end of the face. A small coal-cutting machine cuts an area of coal at this end and provides a 'buttock' or 'start' for the big machine to jib into.

stage loader
This is the control panel in the main gate. It carries the iron scraper chain which takes the coal on to the gate belt and out of the pit. It has all the controls and telephones and display units which monitor 'lock-outs' on the face, etc.

stall
The underground face unit with adjoining gates.

stoneman
See ripper.

tail gate
This is the supply tunnel down which all materials used in the operation of the face pass. Half of the men travel down this gate also, and usually the air passes out of this gate after having been down the main gate and the face.

water note
In some parts of England where men are working in excessive water, i.e. above their knees or getting drenched from the top, they are allowed to work a shorter shift. Either an hour or half an hour is knocked off their shifts according to the conditions they work in. They are issued

with a note from a mine official saying that they are riding up the pit early because of wet conditions. This is called a water note.

NOTES AND ACKNOWLEDGMENTS

Notes
1 See: J. Krieger, *Undermining Capitalism: State Ownership and the Dialectic of Control in the British Coal Industry*, Princeton, Princeton University Press, forthcoming.
2 Paddies have just been fitted up with short-range radios at Hatfield, so speed in contacting them to get in to accidents should now be better
3 Like the Saltley battle mentioned earlier, the struggle to and fro is hard to pin on a card. In June 1982, the Labour Party NEC voted to expel the Militant Tendency, which is assumed by all to be the first shots of a right-wing 'Stalingrad' against all left forces. This is not to say that the process to the left didn't get the thrust from 1972 – but perhaps the acceptance of the incentive scheme and the formation of the SDP has given a counter-thrust to the right. Certainly the industrial mood of the miners will continue to have dramatic effects upon not only their own political consciousness but the colour of the formal political labour apparatus, at the current time still the Labour Party.

Acknowledgments
Grateful acknowledgment is made to the following for permission to reprint previously published material: *Working Papers*: excerpts from 'Britain: Phased Out by Phase Three?' by Joel Krieger (March-April 1978). Copyright © Trusteeship Institute. *Radical America*: excerpts from 'The Miners on Strike'

by David Douglass 8:5 (September-October 1974), *Radical America*, 38 Union Square, Somerville, Ma. 02143, USA.
We also wish to thank Alice Leonard for extremely able editorial work in the preparation of the manuscript and Carol Gardiner, our editor at Routledge & Kegan Paul, for her support and assistance throughout. To Jonah Petchesky we owe a very special thanks.